AuthorHouse™
1663 Liberty Drive
Bloomington, IN 47403
www.authorhouse.com
Phone: 833-262-8899

Because of the dynamic nature of the Internet, any web addresses or links contained in this book may have changed since publication and may no longer be valid. The views expressed in this work are solely those of the author and do not necessarily reflect the views of the publisher, and the publisher hereby disclaims any responsibility for them.

Any people depicted in stock imagery provided by Getty Images are models, and such images are being used for illustrative purposes only.
Certain stock imagery © Getty Images.

This book is printed on acid-free paper.

ISBN: 978-1-7283-4630-4 (sc)
978-1-7283-4631-1 (e)

Print information available on the last page.

Published by AuthorHouse 09/23/2020

authorHOUSE®

SPACE LEADERSHIP
Learning from the International Space Station

Sergey Ryazanskiy | **Pierre Casse** | **Andrey Shapenko**

Contents

Acknowledgments...vii

Foreword ix

About the Authors..xv

   1   Introduction xvii

Chapter 1 – PREPARATION: (Getting Ready for Space Leadership).....................1

   2   An Unexpected Dream 2

   3   A Dream Almost Comes True.. 20

   4   Taking Stock of Your Leadership Learning... 38

Chapter 2 – LIFE IN SPACE (Leading in Zero Gravity) 47

   5   Living in Zero Gravity 48

   6   Taking Stock of Your Leadership Learning... 69

Chapter 3 – MANAGING RE-ENTRY (The Challenge of Coming Back)...........80

   7   Life Back on Earth.. 81

   8   Managing Re-Entry 97

Chapter 4 – THE SPACE LEADERSHIP COMPETENCY MODEL........................105

   9   The Profile of the Space Leader.. 106

   10 Assessment of the Model.. 107

Acknowledgments

The leadership model presented in this book was devised and is described by **Eoin Banahan**, an organizational development consultant and Visiting Associate Professor in Change Management at the Audencia Business School, Nantes, France. It is based on a series of interviews with the astronaut Sergey Ryazanskiy, and written by Professor Pierre Casse and Professor Andrey Shapenko.

I would like to express my deep gratitude to everyone with whom I shared an interesting yet complicated part of my life, training for space missions and the missions themselves - my colleagues.

None of the missions would have happened without the huge team of instructors, who were not just our teachers but also mentors, giving their all to prepare us for the challenges ahead.

Thanks to my friends for your support and to my family – for patience.

Sergey Ryazanskiy

Foreword

Andrei Sharonov, President, Moscow School of Management SKOLKOVO

Space leadership: Who can resist such a title?

This is a very special book, which is not quite in line with what our business school is doing (we are not in space … yet) and yet Sergey Ryazanskiy has been one of our most successful guest speakers. He has been and still is, very passionate about his leadership experience in the ISS and is very good at sharing some of his key leadership learning with our earthy executives.

Some of the discussions he managed around leading an international team in a very challenging and close environment, led to an eye-opening experience for many of our seminar participants.

They loved the exposure to the extreme esoteric situations and yet they were rich in learning. Just take a look at the photographs taken by Sergey and listen to some of the critical events that he had been exposed to and successfully conducted to a very lively exchange of ideas.

It is not every day that one has the opportunity to meet with somebody who dared to go up in space and lead a group of professionals from, basically, scratch. Space exploration is adventurous. Space leadership is unique!

I like to believe that our school of management is at the forefront of what is happening in leadership, not only in Russia, but also in the world. We value curiosity to the utmost and we are also convinced that space exploration is a world trend that executives cannot ignore. We are and will all be impacted (from our existing generations to those to come) by what is happening out there in the wide and open space.

This book attempts to summarize key-learning and unique experiences from a Russian astronaut, that has been confronted with the challenge of leading an international team of experts, living and working together in a confined environment. He was in charge, but had no formal authority over his space partners. He had to make sure that everything went on smoothly, but he could not order anybody around as he pleased. He was responsible for the cohesion of the group and he could only rely on his ability to convince others. His task was, in summarized words, to lead a group of leaders!

This book attempts not only to share some leadership skills and facts, but to do so in a much more ambitious (and yet very modest and reserved) way in order to pull some of the leadership pieces together and to offer a kind of a model that all leaders can get inspiration from:

1. Preparation for a leadership role
2. Being a leader and facing unexpected challenges
3. Giving up the leadership role for something less exciting
4. Up and down, if I may use an easy metaphor.

It offers the reader a unique opportunity to reflect on what they can get from the International Space Station experience (through given exercises) and to challenge some of the reader's assumptions about leadership.

Quite a different reference point to reflect on leadership in today's world, I would say.

This book is a serious entrepreneurial and daring project full of questions. This is precisely what the Skolkovo Professors, Andrey Shapenko and Pierre Casse, who teamed up with Sergey, wanted. A book in motion …

Professor K.Bardach, Former Associate Dean for Executive Education. The Kellogg School of Management. Northwestern University, Chicago.

In almost all cultures, the job of an astronaut is among the most exciting and revered jobs or professions. Especially during the heyday of space exploration, if you were to ask a child what they wanted to be when they grew up, the answer would often be: "astronaut" – if not fireman, policemen or cowboy. And increasingly, the response was gender neutral. Yes, girls, too, often expressed interest in becoming an astronaut.

So, the role of astronaut still captures the imagination of many – especially the young.

Why is that? Why is being an astronaut so captivating? The answer likely stems from the realization that, to be selected for astronaut training, one has to have "The Right Stuff," which is composed of values, personal characteristics, and competencies that are aspirational for most of us. These include: high intelligence, determination, superb reaction time, willingness to accept high risk, highly developed problem-solving skills, patriotism, and likely, a good sense of humor.

In their book, Space Leadership, Ryazanskiy, Casse, and Shapenko, look at space exploration not from the U.S. perspective that we are used to, but rather that of the former Soviet Union, now Russian, perspective. But interestingly, the authors show quite convincingly, especially when referencing activities on the International Space Station (ISS), that these characteristics cut across geopolitical boundaries.

The book is based on the experiences of Sergey Ryazanskiy, one of the three co-authors who himself is a Russian astronaut.

By intent and design, the book does not explore the range of technical skills involved with being an astronaut, the authors state, "We do not underestimate the importance of such skills since they give the space leader credibility, an important source of the space leader's power to lead." Instead, the book emphasizes the core competencies required to be an effective space LEADER. In particular, the authors focus on the space leader's self-management and collaboration skills, which are more concerned with how the leader exercises power.

The implied argument is that if one can identify and effectively implement the core skills required to lead under the uniquely demanding and high-risk circumstances of space exploration, it is possible to apply these same skill-set to a broad range of leadership roles and responsibilities back on Earth with equally good results. Having read the book carefully, I find this to be a convincing argument.

The book is divided into three chapters entitled sequentially:
(1) Preparation – Getting Ready for Space Leadership;
(2) Life in Space – Living in Zero Gravity; and
(3) Managing Re-Entry – The Challenge of Coming Back.

Each chapter contains pieces of the heroic story of Sergey Ryazanskiy, who obtained a PhD in Biology from Moscow State University, and soon after followed research work on weightlessness on the human organism at the University of California, Los Angeles, USA.

He was soon offered the chance to apply and be accepted to the Russian Astronaut Corps. Ultimately, Sergey became first scientist to command a space crew. It was during his second mission that occurred in 2017.

From these fascinating and engaging biographical vignettes, the authors extract and explore the fundamental characteristics, the developmental characteristics, and the critical skills necessary for effective "Space Leadership." The two sets of characteristics and the critical skills are shown as a diagram in the introduction of the book and thus serve as a template for the analysis and discussion that follows. At the same time, we learn and become increasingly riveted and awed by the recount of the actual opportunities and challenges Sergey faced in his leadership role as we proceed through the three chapters.

In addition, inserts are placed throughout the narrative to highlight and explain the connection between each characteristic or critical skill, and Sergey's responsibilities and behavior as a Space Leader. I found these inserts extremely valuable as a means of highlighting particular key behaviors and of connecting the leadership concept to its effective application.

Besides the layout and discussion of the leadership attributes, characteristics and the engaging story about Sergey, whom we come to like, admire and respect as both a leader and a person, the concluding section to each chapter makes the book truly worthwhile for both students and experienced practitioners of leadership. At the end of Chapter 1 there are a series of insightful questions which provide the reader an opportunity to evaluate and assess what has been learned and retained from the section just covered.

At the conclusion of Chapter 2 there is a self-assessment of the six leadership skills highlighted in the Competency Model both in generic form, and as they relate to the reader's improving personal performance. They also have a fun exercise on leading creatively, called "The Small Train Exercise" followed by a second exercise, in the form of a short case, on "The Power of Priorities: Decision Time!". Perhaps the best debriefing at the conclusion of Chapter 2 is entitled, "Information Management: Are you on the top of what's happening?". To set the context for these debriefings, self-reflections, and exercises, the authors write:

"Learning must adapt, invent and experiment with different behaviors to maintain the status quo and succeed. But is it enough when, in a world that is changing fast, we know that most of the things we have learned today will prove obsolete tomorrow? More than ever, what we learn must be questioned and put into perspective. The following self-reflection exercises will give you the opportunity to explore what lies beyond learning."

Finally, at the close of Chapter 3 there is an exercise of both assessment and developing a concrete agenda for personal change.

Taken together, these exercises, which provoke self-assessment, reflection, and discovery, are among the best I have seen anywhere else.

Like space travel itself, the book provides a compelling story about exploration, discovery, learning, application, evaluation and improvement. It is useful for neophytes to leadership like Sergey, when he entered astronaut training as well as for experienced professionals who want to improve their understanding and practice of leadership.

I have no doubt that you, the reader, will be "wowed" and captivated as you read through this engaging and valuable book about leadership.

About the Authors

Sergey Ryazanskiy

Sergey Ryazanskiy is a motivational speaker. His lectures focus on leadership, motivation, team building and working in stressful conditions and draw on his wide range of unique experiences which include being selected and trained as an astronaut, living and working in multi-cultural teams on the International Space Station, performing spacewalks and dealing with crises.

Sergey graduated from the Moscow State University in 1996 with a major in "Biochemistry". In 2003, as a result of his research, he was enlisted as a candidate in the Russian Astronaut Corps.

In 2013 during his first space mission he took the Olympic torch of the Sochi Winter Olympics into outer space. In the same year he performed one of the longest outer space missions in the history of the Soviet and Russian astronautics that lasted 8 hours 7 minutes.

Sergey is also a public figure and since 2016 he has been Chairman of the Russian movement of schoolchildren.

In 2017 during his second mission Sergey became the first scientist in history to be assigned as spaceship commander. He spent a total of 306 days in space and made a total of 4 Space walks.

Sergey has already published 3 books in Russian: "Amazing Earth" and "Amazing Earth. Planet of 1000 colors" which compile best pictures of our planet made from the International Space Station and "How to hammer a nail in space" which answers most of the popular questions about astronautics.

Sergey is a frequent lecturer at the Moscow School of Management Skolkovo

Pierre Casse

Pierre Casse is Professor of Leadership at the Moscow School of Management Skolkovo and holds the leadership chair at the Business School (IEDC) of Bled (Slovenia). He was a visiting Professor at the Kellogg School of Management (Chicago) and the IAE of Aix en Provence (France). Former Dean at the Berlin School of Creative Leadership (Germany) as well as Professor at the International Institute of Management (IMD) in Lausanne, Switzerland. He is a consultant to several multinationals and has published more than 15 books on leadership and negotiation.

Andrey Shapenko

Andrey Shapenko is Associate Professor and Academic Director of MBA Program at the Moscow School of Management Skolkovo. His teaching and research is focused on leadership development and organisational behavior in the context of Russian management culture. Andrey is a certified executive coach and is the author of award-winning teaching cases that are taught at many Russian and international business schools. He is the winner of the prestigious EFMD Case Writing Competition 2016 and a finalist of CEEMAN Case Writing Competition 2017.

Andrey's opinion articles have been published by leading Russian business media (Harvard Business Review, Forbes, RBC, Republic, Inc. and other); and he has been a keynote speaker at multiple conferences in Russia and abroad, including TEDx. Andrey holds an MBA from IMD business school (Switzerland) and a PhD in Economics from Gubkin Russian State University of Oil and Gas (Russia).

I Introduction

Based on the career experiences of Sergey Ryazanskiy, one of Russia's most experienced astronauts, this book sets out to identify the key characteristics and skills required for leading successful missions in space. Following this assessment, a leadership competency model is defined which will then provide a benchmark of successful leadership from which leaders in the wider world can learn. From space to earth leadership!

The Space Leadership Competency Model has three components:

- Fundamental characteristics
- Developmental characteristics
- Critical skills

These components are depicted in the following diagram.

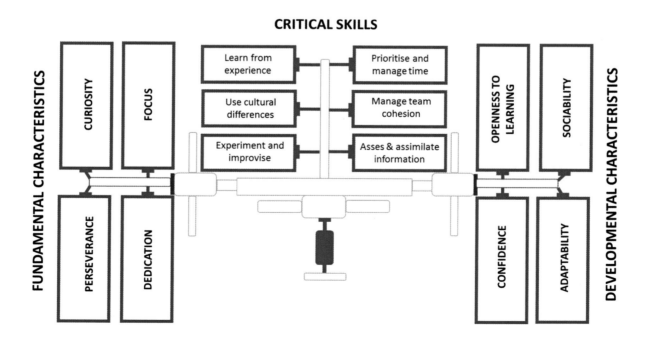

CRITICAL SKILLS

FUNDAMENTAL CHARACTERISTICS

CURIOSITY
FOCUS
PERSEVERANCE
DEDICATION

Learn from experience
Prioritise and manage time
Use cultural differences
Manage team cohesion
Experiment and improvise
Asses & assimilate information

OPENNESS TO LEARNING
SOCIABILITY
CONFIDENCE
ADAPTABILITY

DEVELOPMENTAL CHARACTERISTICS

As Sergey's experiences will demonstrate, leading a space mission involves a complex set of challenges which, while having the potential to make a profound contribution to our understanding of life on Earth and the development of our species into the future, requires a breath-taking range of knowledge and skills, enormous resource investment, and significant risk-taking involving the lives of crew members.

Since 1967, there have been a total of 18 orbital space flight fatalities, of which 4 were Russian, 13 American and 1 Israeli. A further 12 fatalities occurred during training and testing between 1961 and 2014. The career of an astronaut is not without danger and requires courage in the face of risks to life and limb. There is much that the study and practice of leadership in general can learn from the experiences of a successful space leader.

If you are someone who is interested in the concept and practice of leadership, or someone who is interested in improving your leadership capabilities, then this book will be of interest to you. The intention is not to try to determine whether you, the reader, would have what it takes to be an astronaut or space leader. Rather, the intention is to provide you, the reader, with a frame of reference of a successful space leader, based on the experiences of a successful astronaut, Sergey Ryzanaskiy, against which you can benchmark and relate to your role as a leader in your team and/or organisation. Once you have done so, you can then consider how you might improve your leadership effectiveness in your day-to-day activities.

The best way to use the book is to read each part in sequential order since each part builds on the one before. You are advised to go through the book with pen/pencil in hand. Complete the assessments and exercises as they occur and build an assessment of your leadership effectiveness as you progress. The purpose of the assessments and exercises is to stimulate your thinking and self-reflection. It is up to you, dear reader, to decide how to use the learnings to improve your leadership effectiveness.

CHAPTER 1

PREPARATION:

Getting Ready for Space Leadership

2 An Unexpected Dream

"I must say, I never dreamed of becoming an astronaut."

This was Sergey Ryazanskiy's response when asked if it had always been his ambition to travel into space. It is surprising to hear such words from an experienced Russian astronaut. They conflict with our vision of an astronaut as someone who as a child first looks up to the stars and decides to dedicate their life to realizing a dream of colonizing far-off worlds. But then Sergey's career trajectory has proved anything but traditional.

From an early age Sergey developed an interest in biology. His parents, both successful professionals, encouraged him to follow his interests. But nobody in the family, least of all Sergey himself, expected where those interests might lead. During his childhood, the family would spend their holidays camping, exploring the outdoors and what nature had to offer. His parents wanted to show their children that there was a life beyond the hustle and bustle of the big city.

It was a time of adventure for Sergey and his sister, a time when they could act out their dreams, as most children do, in a fantasy world. They could imagine they were intrepid explorers, living a life communing with nature. As Sergey recalls:

> In the forest, you could let your imagination run wild. One day, you could be a famous botanist, exploring the amazing variety of plants and trees around you and on another you could be a biologist, searching for new species which man had yet to discover.

It was a time of great excitement, a sense of freedom from the routine of city life.

Sergey's parents had nothing to do with space research. However, involvement in space exploration was not a new subject to the family, since Sergey's grandfather had been a key figure in the early days of the Russian space program. In those early days, the program was coordinated by a council of six chief engineers, among them Sergey's grandfather. He was responsible for the design of telemetry and telecontrol systems, and he was a well-known, well-respected, and dedicated engineer. Unfortunately, Sergey was too young when his grandfather died to remember him. Perhaps being the grandson of a founding father of the Russian space program was an advantage, but Sergey never felt that his grandfather's career afforded him special privilege. If anything, he felt the opposite, since he would face challenges most astronauts before him never had to face.

Sergey pursued his interest in science, graduating from Moscow State University in 1996 with a degree in biology. His dream at that time was to become a research scientist in a subject he had been interested in since those early days camping with the family, and it was well on its way to realization. Following graduation, Sergey was given the opportunity to work on a research project at the University of California, Los Angeles (UCLA). He jumped at the chance with great excitement.

The study focused on the effects of weightlessness in space on living organisms. The subject fascinated him. He welcomed the opportunity to experience life in another culture, spend time at a renowned university, and meet academics from around the world.

On returning to Russia, Sergey continued his research but now focused on the effects of weightlessness on the human organism. He was becoming more focused on space-related research and was making contacts with others involved in activities and projects on the periphery of the Russian space program. His career was well underway. He looked forward to a bright future in which he could satisfy his interest in scientific research with particular focus on biology.

It was at this time that Sergey was given the opportunity to apply for admission to the Astronaut Corps. The Russian Academy of Sciences was looking to recruit civilian scientists into the space program, and Sergey had the good fortune to be well-positioned for consideration.

It had been three years since he graduated from Moscow State University, and during that time Sergey had been to the United States, pursued some business interests, become involved in some interesting research projects, and applied for the astronaut corps. He had always wanted to be involved in research and had planned his career ambitions with this in mind—but he had expected to fulfill his research ambitions on Earth. Now, however, it seemed that he might pursue his research interests in space.

His application was approved, and Sergey found himself in a new environment: a civilian in a military complex, surrounded by people who were used to living in a highly regimented culture and made it clear that they doubted a civilian would prove up to the challenge. Although Russian space exploration became a civilian program in 2009, the military culture in which astronauts were trained persisted. Sergey felt like a stranger in a strange land from the very outset. Nevertheless, he welcomed the challenge. He could see that now he would walk a new path into the future, one which would open up new learning opportunities.

Training to Be an Astronaut

The training was intensive but provided him with the opportunity to widen his horizons, learn new skills, and explore new interests he never knew he had. Candidates applying for astronaut training need to show an open and inquiring mind and the ability to connect with a wide range of people from different disciplines. The modern astronaut has to be able to find common ground with a large number of research and engineering personnel that he has to represent onboard the spaceship. A lot of people depend on him, and he can't let them down. The astronaut must have a knowledge of many scientific disciplines and carry out research and experiments in a broad range of application areas—he must be a jack-of-all-trades.

As a civilian in a predominantly military environment, Sergey quickly learned the critical importance of fitting in. It's important to socialize and build personal relationships outside the work place. At work, the pressure is on, and people are expecting you to fail, so you must try extra hard to be the best. Sergey had little problem with physical training and the stamina required, but the engineering training also involved studying spaceship systems, ballistics, and other technical domains, and was very tough. He attended lectures during the day, but had to follow up with many hours of concentrated study to try to get familiar with and retain an enormous amount of information.

Basic space training takes two years, during which the candidate is required to take 150 examinations and tests of various kinds. A candidate's progress is continually assessed and graded. Not everyone will succeed through basic training and become an astronaut. The final challenge prior to graduation, is an oral exam lasting five hours, fielding questions from a panel of experts on a wide range of subjects.

The candidate astronaut is expected to be in top physical condition and is required to study a dizzying array of subjects, including engineering, space biology, chemistry, physics, computer science, astronomy, aeronautics, piloting aircraft, survival skills, and languages. The astronaut also has to have a rudimentary knowledge of medicine, since it may become necessary to provide medical attention to a crew member during the mission.

It's a challenge having to absorb so much information. But when you sketch out your subject and break it down into a conceptual model, you realize that actually you only need to remember a small amount of critical information. The importance lies in where to look for the answer when a question arises You need to know whom to ask when you need a solution to a particular problem.. The crew of the space station is the top of the pyramid, but there are teams of dedicated professionals in mission control based in Houston and Moscow, who have developed the systems and receive information and who are monitoring the telemetry around the clock.

The astronaut needs to have immediate recall of the most critical information required for carrying out day-to-day tasks, and should situations arise that require more detailed information, the astronaut must know exactly where to access it.

Once the candidate has passed the exams, he becomes a qualified astronaut. In the Russian system, there are two qualifications, test and research. Unlike the research astronaut, the test astronaut is a specialist with an engineering background that includes military and aviation training. The test astronaut is eligible to become a flight engineer or space commander. The research astronaut engages in research and does not usually have an engineering background. Sergey, as a biologist, first qualified as a research astronaut and then progressed to the group training stage. He hoped that he would soon be assigned to a crew and begin preparations for a flight to the international space station.

The Unexpected Happens

Then in 2003, a tragedy occurred in the space industry. The US space shuttle, *Columbia* disintegrated during re-entry on its twenty-eighth mission, and all seven crew members were killed. As a result, the Americans immediately suspended all plans for future space flights, and the future of the space program in the US was thrown into doubt. However, in 2005, eager to continue with their space research program, the Americans negotiated a long-term contract with the Russian state corporation Roscomos to buy out all the seats for researchers on their space program.

There are three seats on the spaceship. The first is for the military commander, who is a highly experienced flight engineer. The second is a seat for the flight engineer with a good engineering background. The third seat is for payload—someone who is taken to the space station to carry out research before returning to Earth. Following the agreement with the Americans, the opportunity for Russian research astronauts to go into space disappeared. Sergey was advised to pursue another career path, because there was little chance that he would ever be chosen for a space flight.

Devastated at the fact that the opportunities for research astronauts to travel to the space station would now be taken by American astronauts, Sergey realized that the only way he would ever have a chance to fly, he would first have to become a test astronaut. He felt confident that he had the ability to qualify as a fight engineer. However, the culture was not very open to the idea that a research biologist, who had just defended his PhD Thesis, should be given the opportunity to prepare for the role of a flight engineer. It was an entirely different discipline requiring a different mind-set and range of skills. So, Sergey faced the reality of the situation and focused his energy and talents on developing his research career. He was assigned to the Mars 500 research project, which began experiments to simulate a flight to Mars.

A Change in Course

The Mars 500 experiment was an interesting experience, looking at the effects of isolation in space, during which Sergey spent 105 days in a confined space. He worked with an international team and the experience taught him, amongst other things, how to build a competent crew. It was an important project, because future finances depended on a productive outcome. During the experiment, Sergey had the opportunity to command three space crews and with that, the project was an interesting experience in for him on how to build and promote effective teamwork.

Sergey learned that whilst a "prescriptive leadership style" works reasonably well in an environment in which team members were inexperienced, it is less effective where team members are extremely experienced, knowledgeable and competent in their field, because these team members are more likely to discuss, assess and criticize your orders and if they felt that the orders given by you were not effective, they will refuse to fulfill them.

The Mars 500 experiment was also interesting in that it showed how routine, living conditions can affect crew members. For instance, during the experiment the crew had a system where team members held watch over on-board systems in rotation. At night one crew member would keep watch while the others slept. The research set out to explore how the human body responded to having to work with a lack of sleep. The results showed that over time, in an environment where there are no windows and the only light available were artificial, the crew member could not tell night from day and the natural body clock was completely disrupted. When the natural routine of a crew member was disturbed, it had a strong effect on everyone's performance and relationships within the crew. Therefore, it is important that the crew commander maintains a steady routine with regular and sufficient sleep patterns. Once the pattern is disrupted, work effectiveness deteriorates.

A New Direction

Following the Mars 500 experiment, the decision was made to allow Sergey to take all the necessary exams to qualify for a space flight. He had already studied the relevant material for qualification as a flight engineer, so he was fast-tracked through to qualification.

In 2009, Sergey was 33 years old when he became the first qualified flight engineer without an engineering qualification. However, he was assigned to a crew who were quite conservative in their views and not convinced that someone without an engineering background could be accepted as a flight engineer. The fact that Sergey had qualified as both a research and test astronaut didn't seem to matter.

We can now pause in Sergey's unfolding story and consider the fundamental characteristics from his experiences in these early stages of his career path towards becoming a space leader.

Fundamental Leadership Characteristics

Sergey did not set out with the intention to be an astronaut. Moreover, unlike many astronauts in Russia and perhaps more so, astronauts in America, he didn't begin his career in the country's defense forces. However, he had a natural curiosity and interest in science. It was something he found intrinsically motivating and, with the encouragement and support of his family, he matured into a focus on fulfilling his career aspirations. But it wasn't easy. As we shall see, it would require focus, dedication and a good deal of perseverance.

Reflecting on Sergey's formative years and the way he approached his career development process, we can identify 4 fundamental characteristics of the space leader. These characteristics will provide the foundation for our space leadership competency model. Let us consider each of the characteristics as they relate to Sergey's experience.

Curiosity

In those early days of childhood, whilst on holiday with his family, although he didn't realize it at the time, Sergey was developing a broad sense of curiosity which was to prove critical in the years ahead.

A deep sense of curiosity is such an important, fundamental characteristic of an astronaut. A desire to know what it must be like to venture into the unknown, explore the universe beyond our experience and raise new unanswered questions, is a motivation for every scientist. But for the astronaut, it is a driving force!

Curiosity is a natural inclination which is necessary for survival in a dynamic environment. From the moment we are born, we embark on a journey to try and make sense of the world in which we live. Curiosity begins as a restless desire for information and experience which, can lead us towards identifying our fundamental interests. It provides the initial impetus or motivation for our enquiring mind.

However, it can also prove counter-productive in that it can result in an aimless distraction that leads us nowhere. For instance, most people have, at one time or another, experienced a situation when, while online you discover some information on a subject of interest. You follow a series of links which gradually leads you away from the subject you began exploring and before long you end up watching videos on YouTube on some unrelated topic. Without realizing it, you have lost a couple of hours with little to show for it.

For curiosity to prove productive, this initial desire for new information must mature into a quest for understanding. The enquiring mind must become more focused and put a significant effort into investing in and exploring the subject, learning to identify pertinent questions and solve related problems.

Therefore, there is a distinction to be made between the two different types of curiosity. There is the initial, restless desire for information and new experience, and there is the need for knowledge and understanding.

Curiosity
1. *The desire for new experience*
2. *The need for understanding*

The restless desire for new experience makes us wonder what life must be like in space when we look up towards the stars; the need for knowledge and understanding is what drives us towards acquiring the expertise necessary to get there. The mechanism by which a broad sense of curiosity matures into a quest for understanding is called focus.

Focus

Some of Sergey's fondest memories are of his family holidays, enjoying the wonders of nature and the outdoors. These early experiences stimulated his interest in living organisms and fuelled his interest in natural sciences. By the time he finished his schooling, he had already decided to pursue the a degree in biology.

When we refer to being focused, we commonly mean that we are thinking about one thing while filtering out irrelevant distractions. Focusing one's attention is a characteristic which develops through education and the learning process thereof.

People focus attention in different ways and for different reasons. For Sergey, the ability to "focus" was to prove, throughout his career, a critical key success factor. All in all, there are three basic categories of focus.

1. Yourself – "The Inner World"

Focus is primarily a heightened sense of awareness. But we must distinguish between "awareness", which relates to an appreciation of what is going on around us, and "self-awareness" which relates to how we feel about both ourselves and our surroundings. To be focused, leaders must be able to read their feelings and intuitions, understand why they feel the way they do, and control, or manage, the way such feelings impact their attitudes and behavior. Perhaps this is what Apple Inc. founder, Steve Jobs, was referring to in his famous speech to the graduating class at Stanford University in 2005[1] when he said:

> "Don't let the noise of other people's opinions drown out your inner voice. And most important, have the courage to follow your heart and intuition. They somehow know already what you truly want to become."

The focused leader learns how to listen to their inner voice and respond to intuition in formulating their attitudes and behavior.

Generally, there are two approaches to the making of decisions.

Firstly, there is the rational approach in which we carefully assess the context within a situation that requires a decision. We identify potential solutions, run a cost/benefit analysis on each, identify trade-offs and decide on balance, the best course of action. This "rational" approach takes time and requires significant effort in processing and managing a lot of information. As a result, the process can lead to procrastination and, in the extreme cases, extinction.

Decision-Making

1. *Through rationality*
2. *Through intuition*

Secondly, there is a much more efficient, (but not necessarily more effective), approach whereby assessing a situation, the body senses something, either positive or negative, and we make a decision on the basis of "what feels right at the time". Our decision is a response to emotion rather than reason. It is important to understand that the bodily signals that influence our rational processes are also derived from cultural influences, since we develop a sense of what behavior is considered acceptable in society.

---

[1] https://www.youtube.com/watch?v=1i9kcBHX2Nw

As an astronaut, and scientist, Sergey is trained to rationalize and draw conclusions on the basis of the evidence available. However, he must also be able to respond to his gut instincts, an important characteristic when working in a highly stressful environment, in which lives are at risk.

2. Other People – "The Interactive World"

Whilst, a heightened awareness of the way we feel about what is happening around us is an essential element of focus, so too is our awareness of other people with whom we interact. This awareness of the fact that others feel differently and independently of us leads to a desire to understand how others feel.

In our competency model of a space leader, as we will see, this awareness and desire to understand how other people feel will mature into emotional intelligence, an element in the "developmental characteristics" which we will consider later.

Empathy

Understanding how others feel and acting accordingly

The focused leader learns to read people's moods and emotions and how to respond appropriately in that specific situation. This is an important characteristic for the space leader who, as we know must be prepared to spend many months on the international space station, isolated from friends and family, in a stressful working environment and in close proximity to colleagues from different cultures and professional disciplines.

In preparing for living and working in space, astronauts must learn to live together in isolation, away from their family and friends. This undeniably affects their mood and as a space leader, Sergey had to be able to appreciate and respond to other people's feelings.

3.   The Context – "The Outer World"

In the beginning of our lives, as young children, we begin with developing our awareness of the world around us, recognizing our fundamental needs for survival. We recognize when we are cold, or hungry and how to express our need for sustenance. We gradually become aware of other people and differences between them. We recognize our parents and the difference between our mother and father. We also begin to recognize that we are different from those around us.

As we mature into young, healthy adults, this awareness deepens into an understanding of ourselves, those around us, and the wider world. This understanding grows and refines through our experiences and continues for the rest of our lives.

> An awareness and understanding of wider context, the relationship between cause and effect in complex systems, and the impact on oneself and others is an important element of focus for the space leader. As we reflect on Sergey's experiences during his career as an astronaut, we can see this process in action as he moves along his path toward becoming a space leader.

Dedication

> Once Sergey had decided to develop his curiosity in biology, he had to accept that if he was to become an expert in the field, then he would have to invest a considerable amount of time and commitment in studying the subject and fulfilling the academic requirements that are necessary to be deemed an expert.

Whilst focus requires concentration on a particular subject or issue, and the ability to identify and disregard irrelevant distractions, dedication requires sustained commitment of resources over time. This distinction is important, because it highlights a critical issue that is often underestimated in the study of leadership.

To illustrate, a person can focus on producing a progress report, and in so doing, screen out issues that might interfere with the completion of the task, but it may not be dedicated to the task at hand. In other words, dedication requires a more prolonged commitment.

It's useful to think of "dedication", by which we mean, "dedicated effort", as a process that results from a decision to achieve a certain goal. Imagine you, the reader, unlike Sergey when he began his career, dream of becoming an astronaut. The process, through which you have to commit yourself, begins with a dream, but then moves on to the question of how your dream can turn into reality. In formulating an effective strategy, the dream or end goal, must be subdivided into performance sub-goals, which can be translated into performance norms, or behavioral standards. This distinction between "end goal" and "performance goal" is important, because the end goal is something over which you have no control.

Think of an athlete who sets her sights on winning a gold medal at the Olympics. She has no control over who actually wins the race, since she is not the only athlete who is training to achieve that outcome. Her competitors are also preparing and their preparation may, in the end, prove more effective. However, she does have control over how she needs to prepare and what that means for her day-to-day routine.

Now, let's consider Sergey in his progress towards becoming an astronaut. For example, let's imagine Sergey decides that in order to make it onto the International Space Station, over the next ten years his training schedule must involve the following:

Performance Sub-goals	Performance Norms
(Strategy) 1. 12 hours per day studying technical subjects 2. 12 hours per week track and road training 3. 24 hours per week rest and recuperation	(Implementation) 1. Commit your life fully to the astronaut training program 2. Two hour road run each day Mon-Sat 3. Sunday is a rest day

Once Sergey has assessed the nature of the challenges involved and decided on what he considers to be an effective strategy, he will then translate that strategy into specific performance or behavioral norms.

As he implements his program of activities, Sergey must keep a close eye on his potential competition, because the effectiveness of their preparations and progress will, amongst other factors, which are beyond his control, influence any adjustments to his strategy and implementation measures as he moves forward.

Dedication requires commitment to a course of action, which is designed to translate a dream into reality. Since the realization of a dream is dependent upon a range of parameters, which lie beyond your control, you must translate the dream into implementable strategies over which you can exercise control.

For Sergey, as we have seen, the dream of going into space was not within his control. However, it soon became clear what he would have to do to qualify as a test engineer, if he was ever going to achieve such a dream. There was no alternative, other than to revise his strategy and dedicate himself to completing the relevant training. However, as we will see, the training required to become a qualified test astronaut takes a long time and is extremely rigorous and for Sergey, it was anything but "an easy ride".

Without dedicated focus over a prolonged length of time and the perseverance to see it through, qualification as a space leader is impossible. So, this brings us to the fourth element in our model of the fundamental characteristics of the space leader – perseverance.

Perseverance

Sergey's journey in preparing for space travel was not a smooth one. He had to overcome many obstacles and although he had many reasons to doubt that he would ever fulfill his evolving ambitions to journey to the International Space Station, he refused to give up.

The fundamental characteristics common amongst people who achieve outstanding success in their professional lives, highlight the distinction between natural talents and the combination of perseverance and passion. It is clear that society in general, and many organizations in particular, consider themselves as engaged in a war for talent, but natural talent is not the predominant indicator of success. The key success factor is "staying power", or perseverance which is a combination of passion, effort and resilience.

Passion

Passion is a powerful and compelling emotion or feeling; an enthusiasm and desire for something in particular. It's become something of a cliché to assert "finding your passion" is a route to success and so we perceive passion as something which is extremely positive and desirable. But, it's worth bearing in mind that it was not always the case. The word "passion" is derived from the Latin word "passio", meaning suffering. Passion drives us forward towards an objective in such a way that we become convinced that once we have that objective, we will find happiness and contentment. However, this expectation is but an illusion, because once we attain said objective, the feeling of contentment is fleeting.

As we will see, when Sergey eventually became aware of his ambition to travel into space, he already had to urge to return to Earth.

This apparent restlessness is something we have inherited through our evolution as we have adapted to surviving in an environment in which food was not always readily available. There were times when we had plenty, but there were also times of drought and famine during which we had to have the impetus to survive.

Therefore, there is a dark side to passion of which we must be aware. Whilst it can drive us toward achievement, it can also be the source of our destruction. We can become so fixed upon the need for achievement that we become addicted to the pursuit. We become blind to everything but our passion and eventually we burn out.

Effort

Passion provides the impetus which fuels effort. Effort is an exertion of physical and/or mental power for a specified purpose.

Resilience

Investing effort is time-consuming and exhausting, it does not always lead to the desired outcomes. From time to time, set-backs are inevitable. Resilience is the ability to recover from adversity and set-backs.

> Following his acceptance as a candidate astronaut, Sergey entered into an environment which had traditionally been a military culture. As a civilian, he was surrounded by people who were convinced that he was not a suitable candidate and were expecting him to fail. Surviving and achieving success in such an adverse environment required a good deal of resilience.

Resilience refers to the attitude the leader adopts in dealing with set-backs and adversity. For Sergey, there was no doubt that the elimination of opportunity for Russian research astronauts following the Challenger disaster was deeply disappointing and without resilience, he would not have adapted as he did.

Sergey's experiences reveal two things about resilience. It is a leadership requirement in the face of adversity, which is a continuous experience and second, it is required in the face of set-back which is a singular event.

In assessing Sergey's approach, we can identify what it means to be resilient. To begin with, whatever the adversity or set-back, the resilient person does not ask the question "Why me?" but asks "What now?" Sergey's response to both the adverse and somewhat hostile environment of the candidate astronaut program, toward someone from a purely academic background, and the setup of his career plans resulting from the Challenger disaster, didn't waste time in looking for someone, some circumstance or event to blame, but instead looked for opportunities to grow and develop his leadership potential. Regardless of the adversity or set-back, Sergey was able to bounce back and continue to make progress toward his goals.

3   A Dream Almost Comes True

Sergey was relieved that he finally graduated from his astronaut training, but as he waited for his next assignment, he became increasingly frustrated. He had done everything he could to qualify for flight, but there was no guarantee that he would be chosen to fly to the International Space Station.

While Sergey waited for the decision that could drastically change his future, he decided to go on vacation, to take time for reflection and relaxation. He was camping in the Crimea when he received the call he was hoping for. It was 2011 and he was assigned to a crew scheduled for departure in 2013.

Training with the Crew

The crew consisted of three people, Oleg Kotov, the commander, Michael Hopkins, an American research astronaut and Sergey, as flight engineer and payload. The crew trained together in different locations – Houston, Cologne and Tokyo. They were trained on the different systems and modules on the space station and they also took survival courses, preparing for possible survival at sea and in severe winter conditions.

One revolution around Earth takes 90 minutes and if the crew were to land at a specific location designated by the rescue teams, they must send a braking impulse at a specific moment, timed to the nearest second during the 90-minute orbit. If they were even one second out, they could land up to 200 kilometers off course. Therefore, the crew had to be prepared for landing anywhere on Earth. The rescue teams guaranteed that they would find the crew wherever they had landed within three days. So, during this period, they need to use basic survival skills and know how to use specialist survival equipment designed for different environmental conditions.

The survival training proved an excellent team-building exercise during which crew members really got to know each other. Sergey learned that in difficult situations you get to see people as they really are, without the veneer of their particular professional expertise. He realized that very often, conflicts that arise between people are caused by the expectation that other person will act the same way as you do in a situation So, when they don't meet your expectation, you take offence and conflict arises. During crew training, Sergey realized that such an approach was counter-productive.

People are different and each person sees each situation differently, and acts according to their own understanding and perceptions. Therefore, you must observe people, try to understand and accept them for the way they are. Moreover, as a member of an international crew, there is also a cultural component. Cultural differences effect our perceptions and understanding of other people and their attitudes and behavior.

One of the challenges in working with crew members from other countries, is coping with the differences in the various technical languages used. Crew members are used to using abbreviations and acronyms to simplify complex technical jargon but each country has their own approach and Sergey and his crew members had to become familiar with all the differences in complex terminology. When communicating with Mission Control, there must be a common language in place. It's an issue that is often overlooked, or at least, taken for granted. There were times when Sergey felt that he had to learn an entirely new language.

After the crew members completed the various training exercises and simulations, they went through post-exercise debriefings. The objective was to assess and review what could be done to improve performance and learn from the experience. Many people, representing a wide range of expertise, turn up for these debriefings and all aspects of the operation were reviewed. Problems were assessed, potential recurrent issues identified, and steps were taken to eliminate potential complications. The whole process is key to identifying potential risks in advance, so that adjustments to the approach and contingencies can be prepared. After a flight has been completed, there is a 2-week de-briefing that follows. Here, the crew reviews everything with all the scientists and developers of the spaceship involved. All systems are assessed and all directors of the scientific experiments are involved. The crew also consults with doctors, psychologists and other experts to ensure that the crew has the right equipment and so on. Adjustments are then made where necessary.

During every training session, there was a psychologist observing and making notes. The psychologist assessed how the crew members worked together. This is important for mission control, because they must understand how the crew members will react and interact if difficulties arise. Sometimes it is more difficult to remove a single crew member than remove the entire crew.

Training for the Unexpected

Everyone involved in the space program tries to plan for potential emergencies. For instance, a situation might arise whereby a crew experiences depressurization following a micro-meteorite strike which ruptures the shell of the space station. This is an example of an emergency which is foreseen and is one which is planned for. In the case of a rupture in the wall of the space station, a predefined procedure is initiated. The commander checks all the crew members and life support systems on board, then the extent of the damage is inspected. If, at this initial stage, the drop rate in pressure is excessive, and it is an immediate threat to the life of the crew, the commander orders the crew to prepare for evacuation. Such an emergency can occur at any time, and when least expected, so all crew members must be prepared to react and implement the necessary procedures immediately. Another type of foreseen emergency which the crew trains for, is in the event of a fire on board the station.

However, it is not possible to plan for every potential eventuality. There is always the unexpected and regardless of the type of emergency, each crew member is trained to anticipate the unexpected. This requires a certain mind-set and this is something which Sergey has learned to apply in his day-to-day life.

An example, before going to the local supermarket to buy provisions for his family, Sergey automatically starts to anticipate all the potential eventualities that might happen and assess potential contingencies that might be required. For instance, what should he do if something happens to the car on the way to the store, or what if the store does not have the items he needs? What options are available to him should events not go as expected. In the case of multiple options, what are the pros and cons of each? For some people this would constitute evidence of paranoia, but it's the inevitable result of intensive training and development of a mind-set which is prepared for the unexpected. This mindset is critical to ensuring that crew members deal with potential emergencies in a clear and calculated way. Emergencies can lead to disaster if members of the crew panic and lose perspective.

Cultural Differences

It was during this aspect of the training that Sergey came to realize that there were a lot of differences in handling emergency situations between representatives of various cultures. For example, Americans tend to automatically refer to pre-defined procedures when an unexpected event arises, which Russians normally "improvise" on. But the effectiveness of such an approach can prove limited, since procedures detailed in the technical manuals are not always effective when situations arise that have not been foreseen during training.

In Star City, the Russian training facility, the instructor runs trainees through standard simulations and when the trainee is fully prepared, the instructor will vary the scenarios to test the crew members' reactions. Sergey recalls that during these sessions, the American crew members would usually reach for the manual and search for an appropriate procedure.

It was an automatic response. But sometimes the crew has to think beyond the manual and respond according to the unique circumstances as they arise. Sergey assumed that this difference in approach was symptomatic of a difference in the approach to training between Russians and Americans. Without saying which approach is right or wrong, the captain needs to understand how a crew member would react in certain situations, appreciate their natural reaction, and handle it efficiently.

When he was the acting commander, if something unexpected happened, Sergey would take control and issue orders, directing crew members to the relevant actions required, assigning tasks in such a way that the roles of each crew member were clear and precise. To do so, Sergey had to have a full and detailed understanding of the strengths and weaknesses of each crew member and the ability to assign roles and responsibilities according to the unique particularities of each situation as they arose. Sergey learned that when situations arise that could not possibly have been foreseen, you must act according to the demands of the situation.

Before embarking on his space flight, Sergey had to organize all his personal affairs on Earth. It is important that each crew member made all the necessary preparations in his/her personal life before departure, because should something happen whilst they were in orbit, they will not be able to do much to address it. Moreover, if the crew member were to worry about what was happening to their family on Earth, their sleep patterns and work performance would have been effected, which could ultimately endanger the mission. As commander, Sergey learned to monitor the crew members' moods and respond appropriately, for the good of the crew as a whole and the success of the mission.

Working with the Crew

Developing the emotional intelligence necessary to monitor the moods of crew members is a skill Sergey learned during training. Crew members tried many ways to raise their spirits. On one occasion, Sergey floated around inside the space station dressed as a hairy monkey. The prank was relayed to mission control and to this day, people are still puzzled as to how the monkey costume ended up on the space station. On another occasion, since one of the crew members was an Italian, the crew arranged a pizza party. They prepared and cooked 4 pizzas - the first pizzas cooked in space - and made a video of the event which they called, "Space Pizza". They had fun and it helped the crew to stay in a positive frame of mind. Such initiatives provided an insight into Sergey's understanding of his role as the spaceship commander.

During training, psychologists emphasized the importance of humor and the ability to laugh at yourself as an effective technique in offsetting the feelings of isolation and boredom that arose from prolonged periods of routine.

Before embarking on the space station, the crew had to be certified for flight. Once the crew was approved, crew members were then quarantined for 2 weeks. This measure was taken to ensure that the crew members did not bring any infections up into to the space station. The period of quarantine was a good time for rest and reflection after all the training and exams the crew had gone through. There was also a back-up crew in quarantine in case of an emergency.. Back-up astronauts arrived for the delivery of the space craft. The rocket is set vertically on the launch pad. The main crew had to remain in quarantine and go through the final preparations to be able to start work immediately when they arrived at the station.

The back-up crew was allowed to leave the facility and, in line with tradition, they went to lay flowers on the memorials in Baikonur. It's a running joke that the back-up crew has to paint the rocket, which is green when it arrives at the launch pad, but white when the main crew arrives in their space suits. Of course, everyone knows that the real reason for the changed color on the rocket is the cooling process it goes through before it is launched.

Traditionally, crews watch a Russian film before launch (White Sun of the Desert, 1970), which provides crew members with the opportunity to spend some time with family who are allowed into the quarantine facility having undergone a medical examination. It's also a tradition for crew members to watch videos from well-wishers, which have been organized by the psychology services. The intention is to lift the crew's spirits and alleviate last minute anxieties. Crew members also listen to certain music before they make their way to the launch pad. The crew suits up and once the commander has confirmed that the crew is ready, they make their way to the rocket and begin pre-launch preparations.

We can now pause once more in Sergey's unfolding story and consider the developmental characteristics from his experiences during his training and preparations for his space flight.

Leadership Developmental Characteristics

From the moment Sergey was accepted as an astronaut candidate, he had to prove himself. At times it seemed that despite his dedication, hard work and perseverance, he would never achieve his objective of journeying into space. He had to develop the confidence in order to keep going.

We can now assess the rest of the training and preparations which Sergey undertook in his journey towards realizing his dream, something which developed gradually during his research career. As we will see, drawing on his experiences, we can extend the leadership fundamental characteristics of our competency model by adding another dimension: the leadership developmental characteristics as illustrated in the diagram.

Reflecting on Sergey's experience in preparing for his dream, we can see that in developing his leadership potential, some further key characteristics were evident. The developmental characteristics are key success factors in the development of the space leader.

Openness to Learning

Sergey understood the importance and necessity of being open to learning because the trainee astronaut must become familiar with a wide range of technical disciplines. He understood that he would have to be able to communicate with a very wide range of experts from different fields and, in a sense, act as their representative when working at the International Space Station on the various research programs. Sergey saw learning as an opportunity to identify and develop new interests and expand to new horizons.

For most of us the thought of what it must take to become an astronaut is daunting. We think that in order to become an astronaut and venture into space, a person must have exceptional learning abilities. As we saw, during his astronaut training, Sergey had to study widely and pass countless exams as well as other assessment tests on a wide range of subjects, many of which he had no prior knowledge.

This openness to learning is a key characteristic of a "dynamic mindset". For those with a dynamic mindset, a person's interests and capabilities, whatever they may be, can be nurtured and developed through a desire for learning and improvement. In other words, one can achieve anything one wants, provided one has the passion and desire for stretching oneself, sticking to it, and persevering especially when the experience is difficult and extremely challenging.

Adaptability

Following the unfortunate tragedy of the space shuttle Challenger in 2003, and agreement reached with the American space program, it seemed certain that Sergey would never get the chance to travel into space as a research astronaut. For many people, this would have proved the end of their journey, but for Sergey, this was not the case. He realized that he would have to adapt to the new reality.

Adaptability is a process of confronting reality, accepting the need for adjustment and looking for the opportunity, Now, let's assess how he did so. Adaptability is essentially a 3-step process:

1. Acceptance

> The tragedy of the Challenger disaster had a profound effect on space research all over the world. It gave rise to a new reality which changed the course of the space program in both the US and in Russia. Sergey had to accept the situation and begin to consider alternatives.

It is common for people to respond to a profound change in the environment, over which they have no control, with disbelief, denial and a refusal to face reality. Such a response, although characteristic of human nature, drains the positive energy and creativity that is so necessary for adaptability. Therefore, in the process of adaptation, acceptance is the first critical step toward finding new opportunities.

2. Assessment

> As his boss at the time advised, Sergey understood that he would have to adjust his career path. He assessed his options. He could have resigned from space research altogether, perhaps become a full time academic at a premier university, but he held onto his interest in space research and looked for related opportunities.

Once the new reality is accepted, then energy and creativity can be directed to assessing strategic options and preparing for adaptation. The new reality does not necessarily mean that the original vision must be abandoned altogether. Perhaps the original dream, or a modified version of it, can still be realized, but through a new approach. The key question to be addressed is: what new opportunities does the new reality give rise to? Change in the environment, whether unexpected or undesired, is a source of new opportunity for those ready to adapt.

3. Adjustment and adaptation

> Sergey's commitment to space research and his decision to adjust to the new reality ensured that he was well-positioned when the opportunity for involvement in the Mars 500 experiment arose.

Once the new reality is accepted and the implications are assessed and understood, the original vision and strategies can be revised accordingly.

Confidence

> As Sergey progressed on his journey, adapting and adjusting to the reality of the situation, and challenges with which he was faced, his confidence grew. With each experience, he built momentum.

Of course, for the astronaut who depends on his fellow crew members, as well as a large team of experts in Mission Control on Earth, confidence in oneself, is as important as confidence in those around you.

Sociability

Sociability, the quality or state of being sociable, refers to a person's willingness to interact, and engage in activities with other people. It is the leader's desire to project warmth and relate to a wide variety of people.

Whilst we are separating out the following two categories to facilitate understanding, we should keep in mind that they are intimately interlinked. In the context of space leadership, sociability is a composite characteristic. So, let's consider the two elements of sociability in more detail.

Emotional Intelligence (EQ)

Much has been written about the importance of emotional intelligence in the concept and practice of leadership. As a result, the concept has become somewhat diluted and there are as many definitions of the issue as there are articles and books written.

In the context of space leadership, emotional intelligence can be defined as the capacity to be aware of, control, and express one's emotions as well as those of others, and manage the impact. This frame of reference will guide us through the assessment of Sergey's experiences as a space leader.

When emotionally charged situations arise in the workplace, they can often result in the development of resentments, discontent and conflicts. For instance, a colleague, for some reason, is in a bad mood which effects the way in which he/she reacts to other people. For the person at the receiving end of an aggressive outburst or sarcastic reaction, it is natural to pass negative judgment on the aggressor, reject them and feel resentful. Over time this resentment can fester and come to dominate and define the relationship. This is typical of a situation in which there is a lack of emotional intelligence by both parties. But, this is not what happens when we consider the way in which Sergey deals with interactions.

When joining the astronaut training program, and dealing with the negative reactions from others who considered him an outsider, Sergey first recognized that such reactions were the result of the context in which he found himself, rather than a reflection of personal animosity towards him. In other words, rather than to pass judgment and develop resentment, he tried understanding what the reasons for the behavior were and then sought to overcome the animosity by building relationships outside of the work environment in order to fit in.

Similarly, during the Mars 500 experiment, Sergey learned how isolation and the pressures and routine of the work environment can affect the moods of the people involved. It's important that the commander continually monitors the moods and emotions of all the crew members and be ready to react if tensions arise which may give rise to resentments etc.

We will see further examples of this expression of emotional intelligence when we come to consider Sergey's experiences on the International Space Station.

Sergey understood that it was quite natural for people who were used to working with candidates from a military background, to doubt his suitability and express their feelings accordingly in the way they behaved towards him. He realized that he needed to express solidarity and build rapport with those who considered him an outsider and so made the effort to connect with people outside of the workplace to form a mutual understanding and respect in order to diffuse the potential for resentment. We will see this process in evidence when we consider Sergey's experiences at the International Space Station where, as mission commander, he came up with some very creative diffusion actions to maintain a positive atmosphere amongst the crew, despite the pressures of routine and isolation.

Cultural Intelligence (CQ)

Cultural intelligence or cultural quotient (CQ) can be understood as the capability to relate and work effectively across cultures. Cultural Intelligence picks up where Emotional Intelligence leaves off, in that EQ is an appreciation of emotions and how they drive attitudes and behavior at the level of the individual, whereas CQ involves an appreciation of shared attitudes and behavioral drivers at the level of the community, or culture. Leadership in cross cultural contexts requires leaders to adopt a multicultural perspective, balance global demands which, at times, might prove contradictory and within multiple cultures, or subcultures, simultaneously.

As the figure below aims to illustrate, the International Space Program, is a complex collaboration comprising many different cultures and sub-cultures.

If we consider the issue of culture at the most fundamental level, we should begin with Sergey himself. As a unique individual, a result of his upbringing, education and experience, Sergey is a culture unto himself. Every astronaut, regardless of background, is different. But he does not stand alone in isolation. He is part of a crew in a training and development program which has its own unique identity. The astronaut training program is part of a much wider Russian Space Program, which processes, procedures and norms differ from the many other research and development programs in society.

As already indicated, the International Space Program is a collaborative effort between many national space programs and each is a complex combination of different cultural influences, each with its own interests, procedures, language, and interpretations of task and process.

4 Taking Stock of Your Leadership Learning

The following exercise is intended to provide you, the reader, with an opportunity to review and assess what learning you have retained from reading about Sergey Ryazanskiy's experiences thus far.

Assess your grasp of "Space Learning" by choosing (A) or (B) for each of the items listed hereunder with respect to what you have read so far. The challenge is to select in each pair what you think is closely related to what was covered, (i.e. the key learnings) in the first chapter of the book.

According to Space Leadership "good" leaders are able to:			

1	A	Keep their options open.	
	B	Know quite early in life what they want to do.	

2	A	Listen to their "inner voices".	
	B	Be careful with their instincts.	

3	A	Imitate great leaders.	
	B	Find their own ways in life.	

4	A	Search for the "right" situation for themselves.	
	B	Experience and enjoy various unexpected events and situations.	

5	A	Learn from previous generations.	
	B	Be careful with what other generations did in the past.	

6	A	Accept what they know.	
	B	Express curiosity in all aspects.	

7	A	Enjoy cultural differences.	
	B	Care only about their own value system.	

8	A	Do what they do well.	
	B	Volunteer for special projects.	

9	A	Push the limits and try something different.	
	B	Apply what worked in the past.	

10	A	Search for new challenges.	
	B	Build on their successes.	

11	A	Keep open to new learning.	
	B	Consolidate what they do well.	

12	A	Know what their good expertise is and stick to it.	
	B	Be a "jack-of-all trades".	

13	A	Fit in, in various environments.	
	B	Identify situations for themselves in which they can excel.	

14	A	Absorb various pieces of information quickly.	
	B	Learn in a steady, step-by-step way.	

15	A	Know what their key leadership weaknesses are.	
	B	Test themselves on an ongoing way.	

| 16 | A | Rely on their personal, verified knowledge. | |
| | B | Know where to go to get some required information. | |

| 17 | A | Realize that everything is important. | |
| | B | Recall any information necessary in a given situation. | |

| 18 | A | Value research very much. | |
| | B | Value operations and results very much. | |

| 19 | A | Trust their own leaders. | |
| | B | Trust their own inner drives. | |

| 20 | A | Be patient and wait for the right opportunity. | |
| | B | Create the opportunities. | |

| 21 | A | Focus their energy and talents. | |
| | B | Try out different things and diversify. | |

| 22 | A | Build and promote effective teamwork. | |
| | B | Believe that individuals make the difference. | |

| 23 | A | Use "prescriptive leadership styles". | |
| | B | Acknowledge that leading leaders requires a different style. | |

| 24 | A | Know that routine work can be necessary and good. | |
| | B | Keep the team members alert and on their toes. | |

| 25 | A | Look for "fast-tracked" ways to move on. | |
| | B | Concentrate on regular and steady improvements. | |

| 26 | A | Realize that success also depends on other people. | |
| | B | Feel totally responsible for their success. | |

| 27 | A | Mobilize their own energy to succeed. | |
| | B | Rely on others to be successful. | |

| 28 | A | Keep focused on one thing at a time. | |
| | B | Explore many avenues at the same time. | |

| 29 | A | Check and cross-check their understandings. | |
| | B | Know when they know and trust their knowledge. | |

| 30 | A | Create stability out of chaos. | |
| | B | Use contradictions for innovation. | |

| 31 | A | Switch and adapt when necessary. | |
| | B | Commit themselves and keep committed. | |

| 32 | A | Handle issues in a rational and logical way. | |
| | B | Trust their intuitions and gut feelings. | |

| 33 | A | Be realistic and down to earth. | |
| | B | Dream about a better life and use fantasies to succeed. | |

| 34 | A | Reject easy rides. | |
| | B | Take advantage of all good gifts of life. | |

| 35 | A | Stick to their challenging ambitions. | |
| | B | Be flexible and never stop reviewing their goals. | |

36	A	Adapt to every situation.	
	B	Anticipate potential issues.	

37	A	Base their decisions on clear-cut pre-defined procedures.	
	B	Improvise and think outside the box.	

38	A	Know their team members and assign jobs according to their competencies.	
	B	Move jobs around according to the demands the situation they are in.	

39	A	Be agile and be ready for the unexpected.	
	B	Avoid surprises through very good preparation and planning	

40	A	Use their sense of humor.	
	B	Create an atmosphere of seriousness and concentration.	

41	A	Know when to relax and celebrate.	
	B	Keep the energy high and constant.	

42	A	Manage the impact of personal characters on the team spirit.	
	B	Free everybody to be himself or herself.	

43	A	Make a clear separation between private and work life.	
	B	Understand that it is only one life with no separation between the professional and personal.	

44	A	Learn by doing.	
	B	Learn by watching others.	

45	A	Combine a strong ego with humility.	
	B	Behave with humility and discretion.	

| 46 | A | Face reality, both the good and the bad. | |
| | B | Use their imagination to impact the situation they are in. | |

| 47 | A | Focus on opportunities not on problems. | |
| | B | See the problems and transform them into opportunities. | |

| 48 | A | Do not give up and stand up in adversity. | |
| | B | Accept deadlocks and examine alternatives. | |

| 49 | A | Care about others. | |
| | B | Lead for great results. | |

| 50 | A | Avoid conflicts that can damage the team spirit. | |
| | B | Use conflicts to enhance creativity. | |

| 51 | A | Avoid quick value judgments. | |
| | B | Assess people's moods and turn animosity into something positive. | |

| 52 | A | Value trust above everything. | |
| | B | Realize that trust is only one condition for team success. | |

| 53 | A | Practice empathy. | |
| | B | Move on as fast as possible with those people ready to go. | |

| 54 | A | Believe that people must control their emotions at work. | |
| | B | Encourage people to express their emotions. | |

| 55 | A | Know that contexts are critical in understanding people. | |
| | B | Focus on people not on the environment. | |

56	A	Realize that pressure in any situation can impact people's behavior.	
	B	Put pressure on people when necessary.	

57	A	Create a team culture as a base for solidarity.	
	B	Respect individual cultural backgrounds.	

58	A	Promote cooperation among the team members.	
	B	Believe in the virtue of competition within the team.	

59	A	Respect power and use it ethically.	
	B	Fight for power: More is better!	

60	A	Impact people though a strong leading by doing style.	
	B	Delegate and empower people.	

Debriefing

Here are the right answers according to the text within the book-chapter 1 and combining A and B). Please circle the items that you have selected:

A Choices									
1	2	5	7	9	10	11	13	14	15
18	21	22	23	25	28	29	34	35	39
40	41	42	44	45	46	47	49	51	53
55	56	57	58	59	Total Circles				

B Choices									
3	4	6	8	12	16	17	19	20	23
26	27	30	31	32	33	36	37	38	43
48	50	52	54	60	Total Circles				

Score "A" + Score "B"	

Assessment

If you have between 40 and 60 "right" selections:

You have retained some of the key learnings from Sergey - well done! Let's continue in our learning journey.

If you have between 20 and 40 "right" selections:

It seems that you have got some good points from the book so far, but you may have missed quite a few interesting learnings. We suggest that you go back to those points you missed and reflect on them. What do these inputs mean?

If you have between 1 and 20 "right" selections:

We suggest that you re-read and reconsider Sergey's experiences and conclusions of his time leading in space in the above chapter.

CHAPTER 2

LIFE IN SPACE

Leading in Zero Gravity

5 Living in Zero Gravity

Acclimatizing

Sergey Ryazanskiy has completed two assignments to the International Space Station in his career as an astronaut, between September 2013 and December 2017. He spent a total of 305 days in space, during which he carried out 4 space walks amounting to almost 28 hours in total. On his first assignment, Sergey was to fill the role of the flight engineer, whilst on the second assignment, he was the designated spaceship commander. On both occasions, his flight to the station took 6 hours.

When the spaceship arrives at the station and enters orbit, the crew goes through all the preparations and procedures for docking, checking all the onboard systems and coordinating procedures with the crew already at the station. For Sergey, this was a time of intense concentration, focused on the job in hand.

The International Space Station is a 60-meter-long tube with different compartments branching off it. There are modules for research and sections for storage and household requirements - a total of 6 nodes in the configuration. The walls of the station are only 1.5mm thick and the station's solar battery has a span of about 100 meters. The result of the international collaboration is an incredible accomplishment, which underlines what our species can achieve when we pool our knowledge, expertise and resources and collaborate for the common good.

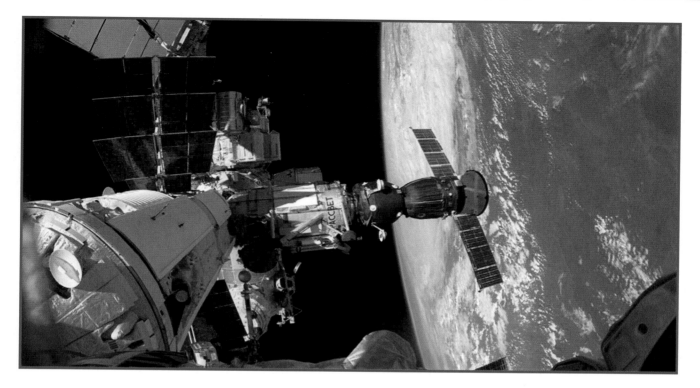

In preparation for living at the station, the astronauts train on a simulator on Earth.

Sergey's first impressions when he entered the simulator for the first time was that it was very small and he wondered how it could possibly accommodate six people with all their equipment. But, when he arrived and settled into the actual station, he realized that the station was actually quite big and could accommodate six people comfortably. It was a three-dimensional space where, in one location, two astronauts could live and work simultaneously – one working against a wall and the other on the ceiling. Both could work independently and without interfering with one other. Moreover, an astronaut can multi-task, working on one assignment using a computer which is fixed to the wall of the station, while simultaneously working on another assignment using the computer which is fixed to the ceiling.

A milestone in the history of the Russian Space Program was reached when Sergey flew to the space station for the first time, as he was the first biologist to fly into space as a flight engineer. He knew that he would have to prove himself. He was well aware that there were a lot of people depending on him and was determined that he would not let anyone down. So, for the next six months of the mission, he was focused, and determined that the mission would be a success.

When the crew arrived at the station, they worked in collaboration with the crew that was already there. This was the handover period. The station had to be serviced and a full research program had to be carried out. There was a lot of work to do.

On his first assignment, Sergey recalls that the first realization to hit him, was that the reality of the station environment was a complete different experience to the one he had trained for back on Earth. In other words, following the months of training and preparations, he assumed that he would know exactly what to expect, but that was not the case at all.

The experience of living in an environment of zero gravity is bizarre to the extreme. For instance, on Earth, as an able-bodied person, you don't give much thought to the mechanics involved in walking from one place to another. Once the decision is made to move, you push off with one foot, place it in front of you and move your body forward as you take your stride. You repeat the process with the other foot, maintaining your balance as you move forward towards your destination. The whole process is so familiar to us that we do it without thinking. But you cannot do this in zero gravity, and if you were to do so, pushing off with one foot, you would simply fly into the walls and ceiling. Unable to direct yourself, you would probably injure yourself and damage the equipment in the process.

On average, it takes about two weeks for the body to adjust to the new environment due to persistent muscle memory and the need to develop a new set of biomechanical skills. Gradually, you get used to the environment and learn how to float and glide through the rooms and corridors within space station.

There are many simple, routine tasks which you do on a regular basis back on Earth, which become much more complex on board the space station. For example, shortly after he arrived for the first time, Sergey was required to take a sample of his blood for analysis, which he was used to doing during the astronaut training, it was a matter of routine. Of course, he had to position himself so that he wouldn't float away, which in itself was rather bizarre, and as usual, he took a test tube, some cotton wool and a device for pricking the skin of his finger. But, as he punctured the skin of his finger, the test tube, cotton wool and globule of blood, floated away in front of his eyes. Despite all the training he had received, the experience was rather disconcerting.

Sergey learned a valuable lesson and from then on out, everything he did was carefully planned out in advance. Whatever the task to be undertaken, before he began the work, Sergey prepared the workplace with Velcro, rubber bands and adhesive tape, just to ensure that the tools and materials he was using did not float away and disappear.

SKILL 1

THE ABILITY TO LEARN FROM EXPERIENCE

Despite the fact that it took about two weeks to acclimatize and adjust, the work routine at the station began immediately. Every moment was meticulously planned out. Sergey was given schedules of tasks, laid out on a daily basis. Each task was timed, minute-by-minute and his performance was continually monitored and assessed by Mission Control on Earth. A certain amount of time was allocated for physical training. It was and is important, and obligatory, that you exercise every day in order to offset muscle degeneration, which is inevitable in a zero-gravity environment. The human body is an amazing machine, but if you don't use the various parts of the body they begin to deteriorate.

The work involved a wide range of assignments, with each crew member's program of activities different to the other. In Sergey's instance, he had a personal program, a national program and an international program of activities. The personal program, and to some extent the national program, involved activities which he carried out on his own. Therefore, some days he would work in isolation with little or no interaction with any of the other crew members. But then there were tasks which required all the crew members in collaboration with Mission Control in Houston, Cologne and Moscow. The international program required a huge international collaborative effort.

Teamwork

There was a strong sense of collective responsibility amongst the crew, such that if a member of the crew made a mistake, everyone would assume joint responsibility. Teamwork was structured in such a way that colleagues supported each other, bearing in mind that those in Mission Control on the ground were also part of the team and that the crew was in constant contact with them. However, the level and intensity of interaction and, to some extent, collaboration between team members depended on the individual crew member's preferred style of work.

During the mission, the crew held a daily press conference (DPC) in the morning. This DPC was devoted to checking the state of affairs on board the station and setting tasks for the day. The DPC provided crew members with the opportunity to clarify any work-related issues and double check schedules and procedures for the day ahead. Then in the evening there was a follow-up DPC to review the work accomplished and prepare for the work to be done the following day.

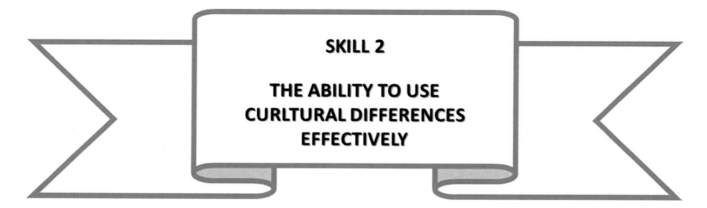

SKILL 2

THE ABILITY TO USE CURLTURAL DIFFERENCES EFFECTIVELY

On one occasion whilst Sergey and his colleague, Oleg Kotov, were on a space-walk, Sergey had to unscrew a bolt. The task procedure was clear. A 13mm spanner, which was sent up to the station from Earth, was to be used, but the spanner was not designed for the work in open space. The bolt in question was situated in a recess and was difficult to reach and therefore they had to make a new tool — attaching the spanner to an extension cord with duct tape. They tried to complete the task with the modified tool but the spanner couldn't fit over the bolt. A reminder that all this was happening in zero gravity, outside the relative comfort of the space station.

They contacted Earth to advise them of the situation but didn't receive an immediate reply. They took look at the documentation and confirmed that the required tool was, indeed, a 13mm spanner. Eventually, they realized that the bolt had been modified, but that Mission Control had forgotten to advise them about it. They couldn't see the modification through the glass visor of the space helmet and therefore the whole task took 40 minutes to complete, much longer than anticipated.

It is inevitable that situations will arise from time to time in which planned tasks do not go according to plan and this can be both frustrating and amusing. Let's say that you have a job to carry out. You prepare your work station and, when ready, begin the task. Suddenly you realize that something you need is missing. You reach for some tool, or other resource, which you placed on the table top just moments before and it's no longer there. You scratch your head in disbelief and look around the immediate area. There is no sign of it and you can feel the frustration rising and as your mood begins to change, you wonder if, perhaps, your colleagues are playing a joke on you or maybe you thought you put it there but are mistaken. You look up ahead of you and catch sight of something floating away in the distance. What's that, you ask yourself, as you push forward. You realize that it's the missing resource, floating away. You reach out in desperation but it only floats further away from you and disappears amongst the equipment fixed to the wall of the station. You begin a frantic search, hoping that the item won't get lodged in some equipment and lead to a bigger problem.

Situations also arise in which equipment fails. For instance, during his first assignment at the station, Sergey and his fellow crew members experienced problems when the energy supply to part of the station failed. In space, there is no convection; warm air doesn't rise nor cold air descend. Therefore, batteries can overheat and explode. Obviously, this is something that Mission Control is very concerned about. It's a major problem which could prove fatal.

As soon as the cooling system fails, larger batteries were turned off and the space station was without power. The crew contacted Houston who has been working on the problem, but they needed time to figure out what had happened and to come up with some procedures to rectify the issue. The crew members convened for a crisis discussion. They talked through all the potential scenarios that they could envisage, sharing opinions as to risks and contingencies.

It was clear that the faulty radiator system would not be easy to fix since it was outside the station. They understood that if they could not fix the problem, there would be a high probability that they would have to initiate lock down procedures and evacuate the station altogether. They went to bed and left Mission Control to come up with a solution. At NASA all the relevant specialists were brought in and an action plan was put together. When the crew awoke the following day, they were given a plan to implement and the American astronauts went out into space and fixed the radiator. It was a period of high tension, with everybody in state of heightened anxiety both at the station and on the ground at Mission Control.

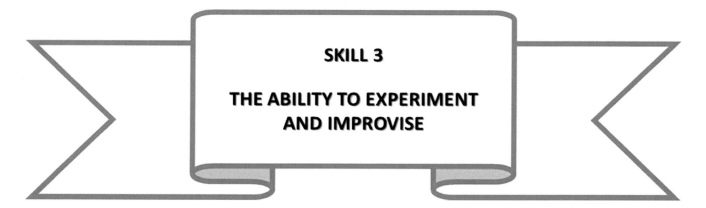

SKILL 3

THE ABILITY TO EXPERIMENT AND IMPROVISE

Team Leader Role

The station commander acts as the voice of the crew after an issue has been discussed and a shared opinion is to be reported to Earth. On the spaceship, the commander has particular responsibilities. He/she has to take a strategic viewpoint on such decisions, as whether to dock at the station or return to Earth. It is a different role compared to other crew members. However, the role of station commander is often seen primarily as a formal role, but this can be misleading. The commander does not control the work of each crew member, he doesn't issue orders and expect crew members to obey without question. Each crew member is a highly trained professional with a detailed program of work which is prepared for him or her by others on Earth. As such, the role of station commander is to unite people, monitor their well-being and offer support if needed. It is more accurate to describe the commander role as a team-building function. The commander is responsible for "socializing" the crew. He ensures that the crew comes together and interacts outside the work routine.

Socializing is very important for uniting the team and maintaining cohesion since the work routine is very demanding and can adversely affect mood and lead to emotional burn-out. Everyone misses their home and family and therefore it is important that crew members have opportunities to disconnect from their work routine and relax. It is crucial to organize events such as birthdays, joint relaxation sessions and progress celebrations as such events bring people together and unite the team. The commander's role is to act as the coordinator of these activities. As Sergey puts it: *"If there is no reason to smile, you need to create one"*. During Sergey's assignments on the station, the crew organized several social events to lighten the mood such as a Halloween party, a Hawaiian celebration, and the Space Pizza party which was already mentioned before.

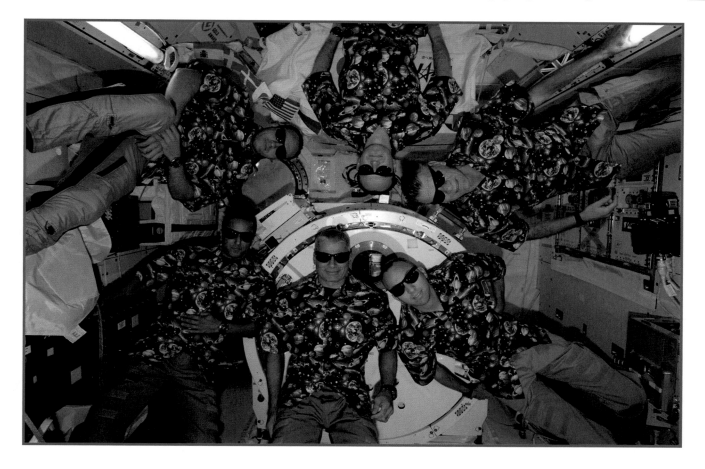

On a Friday, the crew would usually organize a gathering in the Russian module. They referred to it as the "The Russian Club". They would gather and listen to music and celebrate the end of the week. Although both Saturday and Sunday were full working days, they celebrated the end of the week on a Friday. They would talk about the latest news back on Earth or anything else of interest. This regular event was a way of recharging the emotional batteries.

It became a routine too, that on a Saturday they would gather in the American module. They used to set up a small projector and canvas screen so that the room was like a cinema. The American module has wide trapdoors and so they pulled elastic bungees across the trap doors to create make-shift beds. It was an amusing sight since they looked like flies caught in a web, all watching the movie together. Each week, in turn, a crew member would choose a film to watch and then they would talk about it afterwards.

Sergey learned that ensuring team success, is not just a question of complying with the rules and procedures, following instructions and fulfilling assignments, although this is important particularly when the safety of the crew is at stake, but rather, a successful team is one which sets its own standards, defines its own procedures and works to its own rules to ensure assignments are completed successfully. In a sense, the successful team is one which generates its own cohesion and positive culture.

SKILL 4

THE ABILITY TO CREATE
AND MANAGE TEAM COHESION

A successful team is one which forms bonds beyond the workplace. In Sergey's case, crew members got to know each other personally. During training, back on Earth, they often went out together with their wives and children. Their families supported each other and this dynamic proved an important force in uniting all members of the crew during their time at the space station.

Managing Conflicts

Sergey has had the experience of training and working with many people, and one thing was clear to him - every crew, and every crew member, was different. Each astronaut has had different experiences during their life and, as a result, had have a different mindset.

When working in a team in a high-pressured environment, it's not surprising that conflicts arise from time to time. On the space station you are isolated from friends and family but in close proximity with the same set of crew members for six months and negative emotions can build up over time. In your normal life back on Earth, if such conflicts arise, you can take a break, go spend time with friends or go to the gym and work off your feelings of frustration. You can even decide to avoid someone with whom you have conflict. But on the space station you can't do any of this; you have to develop other ways to cope.

It is natural that something, or someone, may get on your nerves at times. Everybody has such experiences from time to time. Someone does something, or says something, which irritates or angers you and you begin to ruminate. As you do so, negative emotions begin to fester and intensify. Like a pressure cooker, as the pressure builds, you need to let off steam before the pressure becomes unbearable, but in space it's a much bigger challenge. You can't express your anger with a fellow crew member since this might have a detrimental effect on the survival of the whole crew. Interpersonal conflicts, if they escalate, can have a negative effect on relationships and the quality of work on the station. So, if something, or someone, angers you, you have to learn to detach and find another way to resolve the conflict since negative emotions will accumulate if left unchecked. In managing conflict on the station, Sergey learned the importance of taking time out and once you have calmed down, you can then talk constructively about what happened and why you got angry.

In order to cope with the stresses and strains of living in close quarters with team members, day-in and day-out, Sergey learned how important it was to have a sense of humor. It is difficult for a person who have only lived on Earth to appreciate what it's like living in such conditions, in which the walls of the space station are only 1.5mm thick. If something happens to damage, or rupture, the walls of the station, the crew will have only seconds to react, or face certain death. In order to be able to react calmly and without panic, you must have a positive approach and therefore, a good sense of humor. Everyone's survival depends on the team. Everyone must support and help each other do what has to be done. Sergey learned the importance of simple, practical things such as when preparing dinner, you do so, not just for yourself, but also for your colleagues.

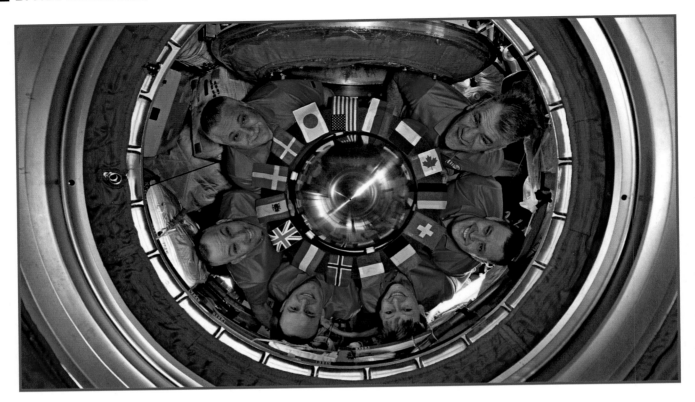

Managing the Work Load

Sergey learned the importance of the ability to prioritize, determine which tasks are most critical and manage his time accordingly. The work schedule included individual assignments and work tasks that he was obliged to do on board. He also had a range of tasks that he wanted to do because they were of particular interest to him, for example, taking photographs or watching sports events on the television. The Space Program's Psychological Support Service would record sporting events and movies, at his request, and upload them to the server so he could watch them when he had the time.

Mission Control would provide him with a schedule of tasks for the day and a detailed agenda. Some tasks had to be done at certain pre-determined times and other tasks required him to be in constant contact with a specialist on the Earth. With other tasks he had more flexibility and so, would arrange his work accordingly, making time for rest and doing the other things he wanted to do.

There are also tasks on the station that require attention of which Mission Control are unaware of. For example, the cargo needs to be sorted or a colleague needs assistance with an assignment. These tasks have to be carried out as they arise or time has to be allocated to them during the course of the day.

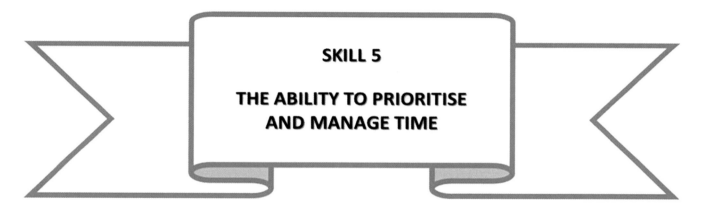

SKILL 5

THE ABILITY TO PRIORITISE AND MANAGE TIME

Sergey found it important to have a hobby, or some activity that he could switch to when he was tired of his work routine. A hobby can be anything, e.g. sport, stamp collecting, television etc. Sergey's hobby and passion was photography.

Whilst photography is a popular activity on Earth which requires a range of technical skills as well as a measure of artistic flair, practicing photography in space has its own unique complexities. It's not just a case of pointing the camera in the right direction and taking the shot. You have to plan out each shot carefully. Set up the camera, find your subject, adjust the camera to the right settings and capture the image. It sounds quite easy, but bear in mind that it only takes a few seconds for you to fly over the object that you want to photograph from the moment you see it.

It's not easy to spot a specific geographical area from space, e.g. Barcelona, Spain. You have to follow from the coastline and observe carefully until you find it. Some landmarks are very difficult to find. On one occasion, Sergey spotted the outline of a farm which from space, looked like a huge guitar. On his first assignment to the space station, he took a picture of it quite by accident. On his second assignment, he spent three months looking for it before he finally located it again. It was in the middle of the South American continent, surrounded by fields. Once he had a reference point, he had no difficulty finding it again.

When taking photographs from space, you have to carefully plan the shot beforehand, timing is absolutely critical and crucial. Sometimes, Sergey would set an alarm clock just to ensure that he had enough time to capture the image from the porthole. Then at the end of the day, he would sort out all the photos taken. Some days he would take as much as a thousand photographs. He would then upload and sort them into folders. On his second assignment, Sergey took about 250 thousand photographs. He had to make a record of what he took, lest he wouldn't be able to identify the images later.

Sergey learned a lot from his first assignment to the space station which was to prove very useful for his second assignment. He realized that although there was a huge array of information with which you had to be familiar with, you didn't need to know everything by heart. In trying to commit everything to memory, you may miss some very valuable information, which you really do need to remember so that you will have no difficulties remembering it when a particular situation arises.

You must try to focus in on the essential information in order not to overload your memory. You need to ensure that you know the key factors in case of critical issues such as emergency procedures or things that Earth, so to speak, cannot help you with. After his first flight, Sergey realized that his knowledge of geography was poor and that it was not something Mission Control could help him with. So, he began to study the subject, trying to improve his knowledge, so that he could more easily locate landmarks from space.

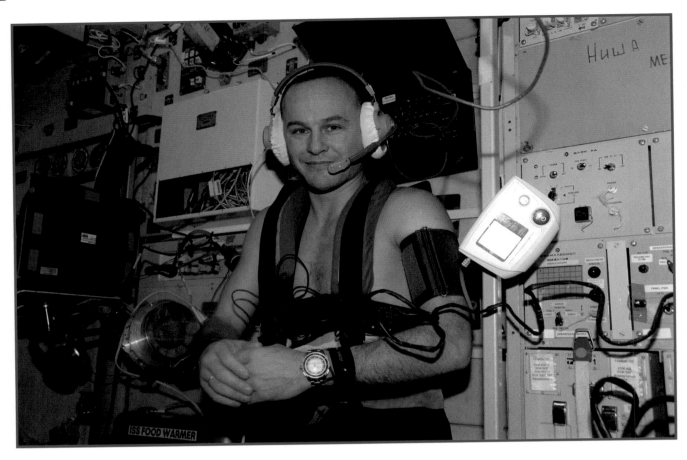

Following his first assignment, Sergey adjusted his approach to learning and concentrated on memorizing what he knew he would have certain use for and to rely on Mission Control for the rest. He also learned the value of teamwork. No matter how much of an expert you are, you can make mistakes, so it's important to have back-up.

Before Returning to Earth

When it was finally time to go home, Sergey thought about all the things he had planned to do while on the station, but never got around to. Before the flight back, he thought about what souvenirs he could take back for family and friends. He took a lot of photographs which made good gifts.

The crew began training for the descent, going over the emergency procedures. As flight commander, Sergey sat down with the flight engineer and went through all the flight documentation. They tried to anticipate any problems and potential emergencies that might occur.

Throughout the preparations, the crew talked with the emergency services on ground, who would be responsible for recovering them after they landed. Here, they were briefed on likely weather conditions. The Soyuz spaceship was reliable, but had little room for cargo. They had to cram a lot of equipment used for the research program, experiments, and personal items into a very small space. Packing is not so easy in a zero-gravity environment. It took a lot of time.

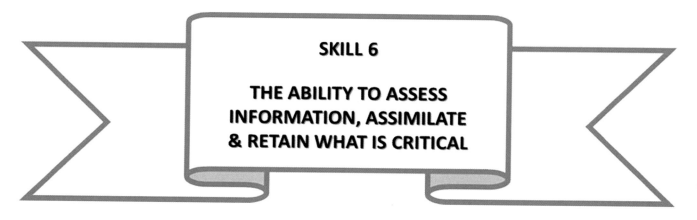

SKILL 6

THE ABILITY TO ASSESS INFORMATION, ASSIMILATE & RETAIN WHAT IS CRITICAL

Eventually the crew boarded the spaceship, ensured all the hatches were secure and moved away from the space station. The crew sent a braking impulse and the ship separated into 3 parts. All 3 parts re-entered the atmosphere separately. Two parts burned up and the crew section fell through the atmosphere towards Earth. The descent was very carefully calculated. On both occasions, Sergey's crew had a smooth descent.

It takes the same length of time to recover on Earth, as it does time spent in space. Sergey found it an amazing feeling when after 6 months of living in zero-gravity conditions, he recovered the normal use of his limbs. At first, he felt dizzy, nauseous and unsteady, but gradually he began to re-adjust to the conditions on Earth. It took several days for the muscles of the body to re-acclimatize. Astronauts have to go through a lot of medical tests and meetings with various specialists after they have descended back to Earth. To help with the re-adjustment, Sergey swam every day and spent time in the sauna, which is good for restoring the circulation in the body.

The crew went through many de-briefing sessions, which started immediately on return while the memory was fresh. They went through all the reports they compiled while at the station and they reviewed what went well and what lessons were to be learned. When he was fully recovered and all the briefing sessions and rehabilitation programs had come to an end, Sergey realized that he just wanted to go back.

6   Taking Stock of Your Leadership Learning

Whilst we do not explore the range of technical skills in the Space Leadership competency model in this book, we do not underestimate the importance of such skills since they give the space leader credibility, an important source of the space leader's power to lead.

In this book, we focus on the space leader's self-management and collaboration skills in the competency model. Whereas the technical skills provide the space leader with a source of power, the self-management and collaborations skills are more concerned with how the leader exercises the power.

In identifying the key skills in our model, we are not saying that these are the only skills required of a space leader, for if we were to include all the skills required, we would have to expand this book into several volumes.

What we are focusing on here, are the "key" skills. So, if we were to list the skills required of a space leader, these would be the 6 most important. Let's now provide you, the reader, with the opportunity to reflect on how the skills might relate to you in your day-to-day work, as a leader.

The following exercises have three objectives

1. Give you, the reader, a chance to review the six skills highlighted in the Space Leader Competency Model.
2. Connect the skills with the leadership requirements, that exist in all teams and organizations in the public and private sectors today.
3. Give you a chance to reflect on your leadership strengths in connection with the skill model and determine what you could do to improve your performance.

Beyond Learning. Are you ready?

Learning means to adapt, invent and experiment with different behaviors to maintain the status quo and succeed. But is it enough when, in a world that is changing fast, we know that most of the things that we have learned today will prove obsolete tomorrow? More than ever, what we learn must be questioned and put into perspective.

The following self-reflection exercises will give you the opportunity to explore what lies beyond learning:

Please assess yourself, on a scale from **0** "not good at all" to **10** "extremely good", for each item listed hereunder.

	As a leader how good are you at … … … …?	SCORE
1	Challenging your existing knowledge.	
2	Making a quick and sound diagnosis in a given situation.	
3	Coming up with new ideas to improve your leadership performance.	
4	Applying and adapting your decisions during implementation.	
5	Learning how to learn.	
6	Checking the impact of your behavior on others.	
7	Changing your mind when something is not working.	
8	Unlearning what worked yesterday but not today.	
9	Turning problems into learning opportunities.	
10	Learning by doing.	
Total		

Debriefing

If your total score is between 80 and 100

Congratulations! There is a good chance you are ready for the challenges of a fast-changing world that requires the ability to challenge our learning on an ongoing basis; learn how to learn, i.e. make a diagnosis, invent, apply, adjust and learn from experience; and, unlearn that which is both obsolete and ineffective.

If your total score is between 20 and 79

It seems that you have some of the basic skills related to learning, and that you are missing some of the key skills required by the new world in the making.

Go back to 3 items for which you scored under 7, and see what you can do to improve. Create a personal agenda for change.

If your score is between 10 and 19

Your situation is almost hopeless. Reflect on where you stand regarding your ability to change and learn. Our advice is to spot what you do well and build on it. Concentrate on your strengths and forget the negative.. We hope that you are a "survivor".

Leading with Cultural Differences in Mind. Can you do it?

Not only must the commander of the station lead a team of leaders, but he must also pay attention to everybody's cultural backgrounds and their impact on the way they work together in space.

Human life is not possible without culture, i.e. the mental programming that we must use to give meaning to everything. Nature provided us with the ability to activate several "cultural codes" that we can use to define the same things. It is a great gift without which we could not survive. However, it is also a major threat to our survival, since we are still destroying each other for the sake of cultural differences.

The following exercise will give you a chance to assess your cross-cultural sensitivity. Read the two models hereunder and then choose the one you feel more comfortable with. This is about leading yourself.

Culture A
Leading is about selecting your goals. You know what you want to do with your life and you organize in such a way that you know what you have to do to get there. You feel responsible for your own life. This model is called **"The staircase model"**.

Culture B
Leading is about being free and enjoying life as it comes, adapting to the requirements of the various situations that you experience. It requires being agile and flexible, and enjoying the positives in life as much as possible while minimizing the impact of the negatives. It requires the ability to detach, so that you don't feel personally responsible since there are so many things that one does not control. This model is called **"The roller coaster model"**.

Debriefing

Now reflect on your choice using the following questions:

- Which one did you select? What are the major things that you like and made you decide for A or B?
- Can you identify the positive and negative sides of both models?

- When do you think that A is more appropriate than B and vice-versa?
- What happens when A people meet with B people?
- What do you think the leader can do to manage the cultural differences effectively? In a team? An organization?
- Cultural differences can create tension and … creativity. Can you come up with some concrete examples?
- How do you assess your own cultural sensitivity (your CQ)?

Leading Creatively: Are You One of Them?

Creativity and speed are absolutely critical in order to perform and enjoy today's world of work. Creativity can be activated at the level of the individual, in that some people are gifted with strong intuition and imagination. It can also be activated at the level of the team as the team members create synergy, or the ability for the team to deliver what the team members cannot produce on their own.

The following exercise will provide you with an opportunity to reflect on your own creativity. Have a look at the exercise below, come back to the description of the problem to be solved and then try to find a solution. We must add that children can solve the problem quickly and easily. What about you?

Problem

Imagine that you can see a train in a station with a circular track stretching out in front of the station. Situated on the track, there are the two wagons, A and B, and at one point the track goes through a tunnel, as depicted in the exercise below. The challenge is to switch A and B by using the train knowing that:

- The train alone can go through the tunnel because the two wagons, A and B, are too wide.
- There is no turning platform on the track or the station.
- The train must be back in the station at the end of the exercise.

The Small Train Exercise

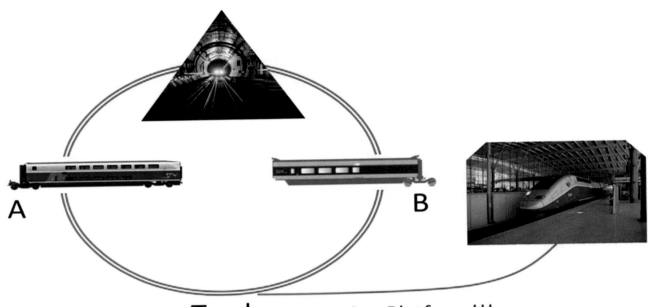

A   B

Track No Turning Platform!!!

This exercise illustrates that:

- Creativity can be very individual (some people have more ideas on how to solve the problem than others).
- There are different types or modes of creativity i.e.
  - Analytical (As long as you have train-A-B in the station the problem cannot be solved).
  - Intuitive; Trials and Errors (To get train-B-A in the station you have to use your intuition (like children) and trials an error (again like children).
- Teamwork is also necessary so that the trials and mistakes can be recorded properly.

Team Cohesion: All Together and Stronger?

Without doubt, teamwork is an important necessity, but we believe that there is a need to re-invent team leadership. Most teams do not deliver results as expected. They suffer from the effects of complacency and compromise. Let's examine some of the common characteristics prohibiting the effectiveness, and efficiency, of a team.

This exercise is based on the TAT (Thematic Apperception Test)

It claims that "leaders' responses, in the narrative they make up about ambiguous pictures of people, reveal their underlying motives, concerns, and the way they see the social world". So, let's try it!

Look at the picture below, take a piece of paper and pen and describe what you see and your understanding of what is happening in the picture.

Debriefing

Now analyze your narrative and see if you have any points in it related to the following team leadership challenges. You can then think about your team leadership strengths.

- Teamwork is about outperforming individuals.
- A minimum of cohesion around the objectives and the strategies are needed to be effective.
- Team members should feel free to stand up for their ideas and fight for them.
- Trust is critical in any team. Emotions should be in the open and dealt with in a positive way.
- Tension, chaos, conflicts and anger are fine as long that is managed properly and are a source of creativity.
- Team leadership is more about empowering people than anything else nowadays
- People should be challenged and enjoy it.
- The team leader should be aware of the "trap of the average" and boost up the team from time to time.
- People's commitment is critical for the success of the team.
- Acceptance and tolerance are conditions for effective cooperation in the team.

The Power of Priorities: Decision Time!

Read the following true story and decide amongst the items listed below, the three most urgent actions that you think the general manager and his top team should take.

The general manager of a sportswear company, based in Ireland, arrived at his office on a Wednesday morning and found a copy of the daily newspaper on his desk. The front page is covering an accident that happened the day before. A young boy, when swimming in the local river, came out of the water with a skin infection and was rushed to the local hospital. He is in serious condition. The article is then accusing the sportswear company of polluting the river with chemicals, which is allegedly the source of the problem. The child's life is in danger because of the unethical behavior of the company's leaders. The General Manager is shocked. He has not been consulted and to his knowledge the company is not leaking any chemicals into the river. He decided to call an emergency meeting of the Executive Team and decide on their priorities i.e. what they should do and in which order.

Please select from the following list of priorities identified by the Executive Committee the three priorities you think they should address and put them in the right order time-wise.

- Check on the boy's health.
- Show that they care about the boy.
- Contact the parents and help them if necessary.
- Check if the company is leaking chemicals into the river.
- Talk to the company employees and workers and reassure them.
- Sue the newspaper.
- Measure the impact of the news on the business.
- Organize a media conference to explain your position.
- Set up an open meeting with the citizens of the town to discuss the issue.
- Use social media to complain about the newspaper initiative.

Debriefing

The three top priorities are:

1. Check if the company is indeed leaking some chemicals in the river. This should be the first move, because without knowing what happened (if anything) the company cannot say or do anything.
2. Show the citizens and also the parents of the young boy that the company cares, (Corporate Social Responsibility is in order here), by issuing a press release, rather than holding a press conference that could lead to some accusation that the company leaders are not quite in a position to address i.e. they don't know whether or not they are leaking chemicals into the river or not.
3. Use social media to show citizens that the company cares about what happened and that they have always been in line with the laws and regulations in that matter.

Information Management: Are you on top of what's happening?

Here is a list of ten points about information management. Can you decide whether you agree or disagree with each item?

Information management is about ...		Agree	Disagree
1	Being well connected		
2	Gathering data constantly.		
3	Checking out the information you get.		
4	Transforming information into knowledge.		

5	Integrating the new pieces of information in your mind.		
6	Challenging what people present to you.		
7	Being careful with what the media present.		
8	Being flexible and open minded.		
9	Questioning the assumptions behind the information you get		
10	Participating actively in the creation of meaningful information		

Debriefing

We suggest that you do the following:

1. Reflect on your answers.
2. Discuss some of the items with your team members.
3. Set up a list of web site providers such as Alibaba, Google, Amazon, and asses your knowledge and use of them.
4. How is your short memory? Do the new technology help you remember things, events and people?
5. How addicted are you to your phone and/or your computer?
6. Do you have a systematic way to collect information, assess it, integrate it into your pool of knowledge and be ready to access when needed?
7. Do you acknowledge that speed is absolutely critical in the management of information?
8. Do you see a need to improve according to your answers to the above questions?

CHAPTER 3

MANAGING RE-ENTRY

The Challenge of Coming Back

7 Life Back on Earth

This chapter is very different from the two previous ones. It is not about the competency model developed on the basis of Sergey's experience. This focuses more on the challenge of re-adjustment which both astronauts and corporate leaders must face once the project which has consumed their talents and energy, comes to an end. It can prove difficult and for some, it can prove depressing.

Whilst in post, the leader is in the spotlight with a hold on the levers of power. The team, organization, and/or society depends on him/her and the weight of responsibility bestows a sense of importance upon the leader. But once the leadership role comes to an end, this sense of destiny melts away and the leader becomes an ordinary citizen, just like everyone else. For some, it can be a difficult reality to accept.

Some examples of such an experience might include the following:

- You are given an important, exciting assignment in a foreign country which you enjoy but the assignment comes to an end and you have to return to your previous position within your company.
- You have been acting as a senior manager but you do not get promoted and must return to your old job.
- You participated in a special project, loved it, but it has left you exhausted. Now it's over and you feel depressed by the lack of challenge at work.
- You were in love for a while and now comes the painful separation.

The question is: how do we cope with what follows a "high" in our life?

To summarize, this chapter is about re-adjusting ourselves to a life that is not quite as emotionally exciting as before. How do we rebound, cope with the down and turn it into a personal opportunity?

Imagine an astronaut like Sergey, up there in the International Space Station, partnering with a group of very special people and experiencing something unique: Flying in space! You go through exhilarating moments which, let's face it, can lead to a kind of "psychological inflation". Then you return back to earth and are confronted with the following situations;

- A great welcome that does not last very long;

- The ordinary, everyday problems of life i.e. a place to live, money to live on;
- The education of your children with whom you have lost touch;
- The reconnection with a spouse who led his/her life while you were away;
- The questions related to our next step (jobs, career…)

What a down-turn! What a challenge! What an opportunity!

Read what Sergey has to say about re-entry. A spontaneous interview to Andrey Shapenko, presented here as it happened. Try to identify in the answers to the questions some of the confusion and ambiguity, that our astronaut friend is indeed facing while back on earth and reflect on what his experience might mean to you.

Andrey Shapenko: Sergey, today we are talking about your return to Earth. Let's first discuss the challenges an astronaut faces when going back to Earth.

Sergey Ryazanskiy: The key challenge is that six months of your life at the station, where you have worked non-stop, are coming to an end, and you need to pack everything up and get ready as if you're going on a long journey. Of course, there are rules to which you must adhere. An example, you cannot leave your personal belongings behind, otherwise the station would turn into a giant warehouse. The crew also has to pack all the experiment results and load them onto the Soyuz to take them back to Earth. The problem, is that the Soyuz capsule has a fairly small cargo compartment, so you need about one and a half to two weeks just to pack everything up nicely, because essentially, you are playing a real life three-dimensional Tetris game, trying to fit everything into the cargo (a tight space in zero gravity), which is initially larger than the compartment. . It's a real challenge with only one, obscure solution. Furthermore, to make things harder, some of our biochemistry test results, had to be brought on board and into the cargo, only hours before launch. Therefore, the biggest challenge was to plan the packing arrangement in such a way that would allow for some last-minute loading of the scientific cargo.

Honestly, this experience is hard to describe; one has to live through it. It's rather difficult to access this cargo compartment, and you feel like a worm stuck in a burrow who has to, at the same time, shove all these packages into a tight space and re-pack them, while talking to the team on Earth. So overall, it is a somewhat peculiar undertaking, the main focus being not to forget anything, because the cargo you're loading is the direct result of your mission, the tremendous amount of work carried out by a large team on Earth and the flight crew on the station—basically, it is everything we flew to space for.

AS: Do you travel back on the same ship that carries the cargo, or on a different vessel?

SR: In 90% of cases, crews fly to and from the station on the same ship. There have been instances when crews had to swap ships, but it comes with additional complications. For instance, you have to drag your seat liners, in which you launch and land, your spacesuits and some packed cargo to another ship.

AS: So, basically, you arrive and dock at the station on a ship, which will sit there until you complete the mission?

SR: Yes, the ship remains docked at the station waiting for the crew to complete the mission. During the six months of the mission, you slowly store things that are ready to be packed in a particular place. Then, since out of the ship's three modules, only one is used to carry both the crew and the cargo, which is quite small, to Earth, you have to load all the things you've accumulated during the mission into this container shortly before the boarding.

AS: How do you prepare for the descent?

SR: This is a very important aspect, because the descent is one of the most dangerous parts of the mission, in addition to being physically demanding. Even though astronauts exercise for two hours every day during their stay at the station, it's impossible to engage every single muscle, and in the preparation for the landing, we have to intentionally train some particular body systems, for example, the cardiovascular system. Moreover, you have to understand that it will be subjectively difficult. It is easier for a regular healthy person, living in the conditions of Earth's gravity, to withstand the G-force during the descent than it is for an astronaut after six months in weightlessness. Unfortunately, our bodies quickly adapt to new surroundings and lose their ability to handle G-force.

The second important aspect in getting ready for the landing is technical training. Since it has been more than 6 months since we last revised all the information on the technical systems, we needed to freshen our memory on some of our own rules and skills, learning technical updates and new incoming information on difficulties from Earth. This preparation takes place about a week before the flight back. We collaborated with the instructors that once prepared us for the launch. They help us to refresh our memories. It was an essential part of the training because both the descent and the landing are two of the most dangerous stages of the flight.

AS: What happens during this refresh session? How did you prepare physically? Did you receive some instructions or did you work out some actions and went through some simulated scenarios?

SR: Preparations take many forms. To start with, there is on-board documentation that you just have to sit and study, line-by-line, step-by-step, reminding yourself what each particular action means, what emergencies can arise at any given moment and the potential consequences. Basically, I went through both the planned course of actions and the so-called "foreseen emergencies".

In addition, we had a computer program simulating descent that allows us to train for the event. There is a manual descent mode for when on-board computers fail and you have to manually pilot the descent with controllers and special engines. These simulators are used to master manual descent and to test-fly in different modes, to see what kind of emergencies and conditions you may encounter. The training in the simulator also refreshes your actual skills and muscle memory of controlling a ship during re-entry into the atmosphere.

AS: What is the most challenging aspect of the descent itself?

SR: The two main challenges of the descent are that, first, everything happens rather quickly, and second, no one has yet missed and flown past the Earth. The most important thing is to lessen the G-force to make it easier on the crew, because on the way down we have to control the flight process, but the descent can be so stressful on your body, that you physically cannot do your job.

AS: What do you feel in these instances? Do you feel excess pressure or lose consciousness?

SR: It's difficult to describe the feeling accurately. First of all, you do feel actual pressure and the force of gravity crushing your insides. If you exhale all the air from your lungs, you cannot then take a deep breath, because it feels as if there is a small elephant sitting on your chest and your muscles simply cannot lift it to let the air in. The weight of the G-force is just too much.

Second, the vestibular apparatus, which you don't use and can't train while in space for half a year, suddenly has to function again, and it doesn't go smoothly. You may experience some vestibular illusions and feel nauseous. It might not happen immediately; people and situations vary; some feel sick after landing. The point is, the vestibular apparatus suffers from not having to work for six months and then suddenly feels all the might of Earth's gravity. You experience all this while you have to report to the command centre, closely monitor all the indicators of the ship, and keep some key elements in mind. Despite the discomfort, you have to be fully engaged, because descent is a critical process, which, unfortunately, cannot be completely automatic.

AS: Do you know of any steps you can take to protect or prepare yourself for the trials of the descent, apart from training?

SR: First and foremost, we are professionals and we understand that we prepared for it and passed all the necessary exams, so, it's nothing new and we can do it. It's also important to remember that each person's G-force sensitivity and susceptibility to vestibular impact is different, so any crew members can lose consciousness or become unable to do their part at any given moment. That's why the entire team is engaged in the process, ready to help out and draw their attention to some details you might miss, because things can get complicated—sweat might cover your eyes to the point where you can't see and you can't even wipe it off, because you're in a glass-protected spacesuit or, the G-force may become so intense that you can't report the flight, so someone else has to pitch in and help if they notice that you are not coping well.

This is what teamwork is all about, when any member can step in to back up the leader. For example, if a rescue aircraft is inquiring on your status, but the commander isn't answering for some reason, it is the responsibility of the flight engineer to communicate that, "This is Borey-2, we are okay, the signal is good, everything is under control, indicators are such and such." Sometimes the signal is weak due to the plasma interference and you can't hear the commands from Earth very well, and everyone's physical conditions can be different. One person might be doing well and be able to answer, while another is fighting for their life. That's why we focus so much on teamwork and try so hard to cultivate good cooperation in the team before the mission.

> **AS: What are the difficulties and challenges that arose upon your return to Earth?**

SR: Once you land, you realize that it was much more comfortable to live in zero gravity. Our bodies easily adapted to diverse conditions, and when we felt at ease, floating in the air without much physical stress, it disengaged some systems and muscles that gradually atrophied. And then suddenly there was all this work for the vestibular apparatus, but for most of us, it recovered rather fast.

I was fortunate enough to handle the flight back and the landing relatively well, but there have been cases when people couldn't get out of bed for three to four days because this ordeal took a toll on their bodies. Despite how much you trained, it is difficult to walk, your back hurts, your knees hurt, your joints hurt, your body has grown unaccustomed to intense physical activities, so you have to get in shape slowly, not pushing your body too hard, because if you rush it, it can lead to serious injuries. So, you need to re-acclimatize to your old life at a slower pace.

There's a mandatory daily fitness regime: swimming, sauna (to restore muscles and blood vessels), and moderate relaxed exercising under medical supervision. This rehabilitation required daily effort and took almost as long as we were in space. Of course, we felt relatively normal in about two months, but then it still took time to get to our former condition.

AS: And you needed two months to get back in shape?

SR: It took two months before I felt like a normal person again. At first, you wake up with ear ache. My ear felt as if someone has crushed it, but in fact I simply turned in my sleep and landed on the other side of your head. What do we normally do on Earth all day? We sit. We sit a lot, especially during meetings. And even when you come in for a medical check-up, the doctor tells us to take a seat. But upon my return, I soon learned that your rear is no longer used to sitting, so gradually it turns into one big bruise, if you don't bring a pillow with you everywhere. It is an example of those muscles that are virtually impossible to exercise in space.

Your joints also get swollen a lot; you rush through a flight of stairs and your knees start hurting in return. Your back hurts, because you stretch out in space—some astronauts grew up to five centimeters—and then when you get back to Earth, you start shrinking back: it happens because your spine expands in weightlessness and then compresses under the Earth's gravity. It is an uncomfortable and rather painful process. That is why swimming is highly recommended—it relieves some of the pressure and lets the body relax.

In addition, the first two weeks are jam-packed with medical examinations; you go from one doctor to another for a full body check-up. On one hand, it has scientific value, — to examine the impact of microgravity, space flight and cosmic radiation. On the other hand, it is necessary to professionally monitor your physiological condition. Unfortunately, space flights are not the healthiest affair as it changes your biochemistry and physiology. After returning, it's not so easy for your body to bounce back.

For instance, during the mission your calcium reserves can become substantially depleted. You can lose up to 7–8% of all calcium in the body. Densitometry performed during the check-up reveals decreased density of the bones, the teeth, and so on. You recover little by little, but not as quickly as you would hope. That is why you need a special diet, medical supervision, and exercising with a personal trainer—all requiring patience and daily effort.

AS: Approximately how many hours per day do you train during these two months?

SR: About two hours every day. The training starts the day after the landing, and at first, it's two hours plus examinations, then two hours of training plus debriefs.

AS: Could you describe the post-flight debriefing?

SR: Two weeks of the medical rehabilitation are followed by two weeks of technical debriefs with all the experts on the ship's and the station's systems and the scientific supervisors in charge of the experiments. Of course, we also conduct mini-debriefings during the mission right after each experiment or repair work, but these two weeks are when we do a full-scale analysis and review the results. "This is what you told us; this is what the telemetry told us, do we understand correctly that this and this happened?" Basically, what we do is, is we sum up the results for the official flight report.

This discussion involves several hundred people. Each day a new group of specialists come to discuss something with you, and you can see in the schedule that simultaneously experts are analyzing the results of biological experiments and physical exercises. Your workout sessions are attended and observed by your trainer, treadmill designers, bicycle designers, people who were in charge of purchasing expanders and doctors who monitored your health. That is, people are carrying out a full comprehensive analysis of the mission—from your diet, personal hygiene, and fitness to mega important experiments. Then during debriefing you might already learn some preliminary results.

At this point you also realize that the work has never stopped and it is going at full speed: "Sergey, during the flight you noted the following flaw in the design; look, we have fixed this and that, and instructed the engineers to do this." This shows that the crews involved in the next mission already used your notes in their preparation and you are just concluding the results: "Yes, I agree with this; it would also be nice to tweak this and that." All this work is aimed at preparing the final mission report and receiving feedback;

, feedback is the fuel of progress and improvement, it is the culmination of the entire flight. We even have an inside joke that while the debriefing and rehabilitation were still happening, the flight has not yet come to an end. It is only when you've recovered that your flight is truly over.

AS: What helps you to adapt and recover faster, both psychologically and physically?

SR: First and foremost, it is the continuous cooperation with the experts, with the crew physician who monitors your condition, the personal trainer who works with you every day and gives you new tasks, new tests, and examines your progress. In addition, after being away from your normal life for half a year, you get a really good boost from reconnecting with your family, spending time with your children, visiting parents, talking to your wife/husband and meeting friends.

This is probably the most important thing in life, it gives meaning to our lives and our work, allowing us to return with a sense of fulfillment and pride for the mission carried out. Then, once you are back, support of your family and friends are the key elements helping you to return to your normal life.

The space flight does not finish at the moment of landing, there are 6 months of daily procedures meant to facilitate body recuperation and re-entry into regular social life.

A major personal task after the flight, is to understand whether you will get assigned to a mission again or if you are back to your previous lifestyle for good. It resembles the situation when one is starting a job in a new company or moving to another country. You have to start from a scratch. In the meantime, the experience you gained in space might be a good start point or baggage that is dragging you down. One has to make a decision: what do you really want, what amount of energy and time do you have and, how does it affects your family

The preparation, flight and rehabilitation takes a minimum of three years. While an astronaut is up there, life goes on, and they take no part in it: he/she does not see their children grow, does not participate in friends' lives, does not support parents who are growing old. It is a challenge of its own upon return to get to know your friends and family again and restore lost connections.

Another issue is that you can't really discuss what was happening to you during these three years, it's hard to imagine for an ordinary person how it is – to be an astronaut. The challenges and routine are too different from those that your family and friends are used to. Misunderstanding might become a serious obstacle during re-entry.

> **AS: When talking about the launch, you said that astronauts have special pre-flight rituals, like watching a particular film. Do you do anything similar, some other rituals, after coming back?**

SR: Yes, we do, and our American colleagues do as well. In addition to the post-flight debriefing, which is done by both the Russians and Americans (sometimes with Japanese and European specialists), there is a kind of a closing event, a ritual or a ceremony of sorts, where we sum up the results of the flight.

This event provides an opportunity for the returning astronauts to thank their instructors. Crew members usually prepare some gifts, for example, collages made out of some beautiful photographs they took in space and present them to the Earth specialists who provided them with valuable support and professional advice on some systems and scientific work, or who encouraged them during tough operations. Simply hearing someone giving you necessary information in a calm voice, can help you cope with the stress and emotions.

This is basically a thanksgiving ceremony, during which you express your gratitude to those who took part in this mission, kept you safe, and provided you with everything you needed. It's a pretty exciting event both in Russia and in America. In the US, the ceremony takes place in the Space Centre Houston near the primary NASA grounds, where our American colleagues film a special commemorative video after the flight, present diplomas and celebrate with their families. They also bring their children to show them that their mum or dad were a part of such important work.

In Russia, we celebrate this occasion by walking through the Cosmonauts Alley, across the entire Star City, and then holding a ceremonial meeting in the Cosmonauts Cultural Centre with awards, flowers, speeches and so on.

AS: Are the families invited?

SR: Of course. Families and children participate in the ceremony, and even regular Star City residents join in—because Star City is a unique place, where each person, even if they currently don't work in the space industry, have either been involved in some capacity in the past or have family members affiliated with this field. One way or another, everyone in this town lives and breathes space travel. The event is open to the public and attracts quite a number of people— instructors, admirers and some aspiring astronauts who have not yet had a chance to fly but who wish to see how it all happens and to feel involved.

AS: You mentioned that this event is an opportunity to thank the people who helped you on this journey. How many people are we talking about?

SR: I would say about several thousand people. There are multiple hundred specialists that constantly surround you—technical staff, doctors, trainers, scientists, instructors, etc. But there are also people whom you recognize solely by voice, as they were the ones who communicated with you from the Mission Control Centre on Earth and provided you with the latest information. There are also the scientists that developed the experiments. In the course of the work, you only see the tip of the iceberg, but you know that this visible group represents a much larger team that designs the equipment or prepares elaborate scientific experiments. That is why you genuinely try to thank as many people as possible.

AS: If I remember correctly, you mentioned in one of the interviews that after the flight, you had a chance to relax in addition to attending debriefings. Would you recommend some particular method: do you go on vacation with your family or alone, do you prefer active vacations or spending time to unwind on the beach?

SR: It all depends on the particular person. Some astronauts are full of energy and need to keep moving. Others prefer to lie down and rest, think about their experience and the future. Once you land, you first go through two health-oriented weeks, then through two weeks of post-flight debriefings. Then the three weeks of the official rehabilitation starts.

This is a purely Russian tradition; Western space agencies do not follow the same protocol. Our astronauts are required to take care of their health. You have twenty-one days that you must spend relaxing. But you do not go alone: this is not vacation time per se—it is recovery time. Your personal trainer and crew physician join you on this trip and have four to five workout sessions a day, including physical therapy and massages. On top of that, if you have some specific health problems, there is a special recovery program for those who need it.

After you are done with the debriefing and rehabilitation, you are free to go on a real vacation, and the type and destination is up to you. Obviously, most astronauts try to go somewhere with their families and spend the time relaxing. However, the specifics vary from person to person: some decide to travel, some sunbathe on the beach and contemplate on what to do next, some just want to spend time with their families, because, unfortunately, during the pre-flight training, the mission, and the first few weeks after the return you can rarely find time for them. So that's basically the first opportunity in a long while to spend some quality time together.

> **AS: You've been to space and back twice. Do you have any pieces of advice for future astronauts on how to quickly adjust and return to their regular lives upon return? What should they definitely do or definitely not do?**

SR: The most important thing is to have some aspirations, some new goals: "I need to recover quickly, so I can go do this …"

Sure, you can rest on your laurels and pat yourself on the back for the job accomplished. But the only way to move forward is to set a new challenge, a new goal: "I need to get back on my feet and prepare for another flight," or "I want to try this, and I need to be ready soon because people are waiting for me." I believe that you need to constantly invent new objectives, both to motivate yourself and to push other people.

Secondly, it's probably pretty obvious, but it bears repeating—it's going to be hard and you'll have to just endure it. It's important to realize that you'll require patience not only while waiting for your flight, but also when you get back—when it gets unbearably difficult and you have to keep pushing forward towards recovery despite the pain and discomfort. Unfortunately, astronauts are really bad at being weak, so when they get back and feel totally incapacitated after being strong and athletic throughout all their lives, it takes a toll on their confidence and emotional well-being. It's easy to gain weight if you don't make an effort to maintain your physique, and easy to let your health deteriorate. So, you have to bite the bullet and keep working to return to your pre-flight condition, or at least achieve 99% of this task.

> **AS: Do you have a third piece of advice to make it a set? Self-motivation, ability to bear, and?**

SR: You probably also need to learn to reflect on where you succeeded and where you failed. I mean this in a more personal sense, a kind of self-reflection, possibly even off-the-record. It's one thing when you analyze things during the debriefing, and then another thing entirely when you can self-examine what you were able to do properly and what you needed to do better, what you enjoyed, and whether it is worth repeating, or maybe it's better to leave it behind and move on to something new, what you should do if you fly again to make your next mission more interesting, more comfortable, and so on, In other words, you need to sum up the results of the flight on your own, regardless of the debriefing conclusions. This helps a lot with deciding on what to do next and how to bring your next undertaking to a new level.

8   Managing Re-Entry

To come back to earth after having experienced some unique events, such as walking in space, can be a difficult turndown experience. From space to earth can be quite a "step back and down"!

However, it can also be an opportunity to reflect on the profoundness and uniqueness of the space experience, to share some learning with other people and also to move on with one's own life, enriched and more meaningful. The main challenge is to accept that life will never be the same again.

As a leader, whether at the level of the team or organization, albeit in a different context or with less intensity, you experience the same process when, for instance:

- You accept an international assignment which is good for your career and you go through a re-entry phase when you come back, all excited and full of ideas about what you could do next. But as soon as you return, you realize that nobody is waiting for you. Life went on without you. The power game continued as normal, despite your absence. The disappointment can run deep and prove taxing. You go back to the old routine, deflated and depressed.
- You have been acting as a senior executive for a while and despite the fact that you have been performing well, you are informed that you are not going to get the job after all. Politics has apparently interfered with your legitimate ambitions. You must return your old job, which now seems so dull and meaningless. You don't think it's fair and have difficulties accepting that decision.
- You were asked to take over a special assignment for the CEO of the company. You took it and it became a challenging and exciting assignment. You felt that you were part of the decision-making process and that you could contribute in an important way to the success of the organization. You regularly met with the Executive Committee, reported on your findings and made some suggestions on how to progress. However, the assignment has now come to an end. You have received the CEO's congratulations and must now return to your regular job. What a disappointing result.

- You have had a great career in a fast-growing organization. You enjoyed your job very much. But now, it's over. It is time for you to retire, rest and enjoy a less stressful life. But you are not ready for that. Re-entry, in the sense that you are moving to a new phase in your life, is very difficult for you. You had such an exciting life before.

Leaders facing a re-entry issue are confronted with three major challenges:

1. Coping with ambiguity
2. Leading change
3. Managing emotions

Let's examine the three key challenges, one at a time, and see what leaders can do to manage re-entry meaningfully and successfully keeping in mind that:

- Each experience is personal and that solutions must be adapted accordingly.
- Only a few "guidelines" are provided and the best ones will be the ones that the leader will invent for himself or herself.
- Re-entry is just a phase in life that never ends.

Ambiguity

We experience ambiguity when things are unclear or when the situation in which we find ourselves is loaded with contradictions or paradoxes. We feel lost and unbalanced: we do not understand what's happening nor why and we don't know how to get ourselves out of the confusion. Uncertainty is what we experience when the ambiguity goes on for a long time and there is some unpredictability regarding the future.

There are three basic ways to cope:

Research in this field shows that there are three behaviors that can help "turn ambiguity into an opportunity". We suggest that you pick the one you feel more strongly about as a reflection exercise. Furthermore, consider the negative side of each option.

Behavior 1.

Withdraw from the situation, at least temporarily so that you can put things into perspective. Take a step back and look at it from a different angle. There are different ways to look at any event. Learn from what's happening. Reflect and grow from the experience.

Negative side of this behavior

To withdraw permanently without learning anything from the experience is called "fleeing".

Behavior 2

Take a risk. Take initiative. Ambiguity is a door opener to exploration and new discoveries. Things are not clear. Go for it then. It is an opportunity to create something on your own.

Negative side of this behavior

Control yourself and don't become overconfident or even aggressive. Pushing back is appropriate as long as it helps you create something good. Don't fall into the trap of believing you are absolutely right and start to blame others for their blindness.

Behavior 3

Analyze the situation. Review its negative and positive sides. Try to get a good understanding of what happened. Focus on the why. Get a good learning from the experience.

Negative side of the behavior

Over-analyzing

Leading Change

Change is constant. We never stop transforming ourselves. Sometimes change is desired and expected. Sometimes it's totally unexpected.

One extremely powerful way to lead change is to use the so-called Objective Self-Examination (OSE) approach to change. It consists in:

- Reviewing your strengths;
- Facing your weaknesses and accepting them;
- Discovering your potential.

The OSE approach can be activated the following ways:

1	**I do it alone**	I reflect on what happened during the day and learn from it.
2	**I get help**	I ask somebody who knows me well, and in whom I trust, to play the role of a "sounding board".
3	**I consult**	I consult with appropriate professionals, such as a coach, to clarify a direction on how I can find my own solutions to the problems I am facing.

Consider the following approach to setting up your own agenda for change.

Agenda for personal Change	
What I am not (Weak points)	
Who I am (strengths	
Whom I could be (Potential	
Actions to be taken	

There is also the classical way to lead change: Decide on what you should stop doing; Start doing; Do differently.

Managing Emotions

Assess yourself for each of the items listed hereunder using the following scale as a guide:

(1 = Not good at all; 5 = Moderately good; 10 = Very good)

	How good are you at ...?	SCORE
1	Coping with stress	
2	Controlling your anger	
3	Transforming your frustration into something positive	
4	Putting things into perspective	
5	Expressing your happiness	
6	Enjoying personal challenges	
7	Trusting yourself	
8	Seeing things from various angles	
9	Energizing yourself when needed	
10	Learning from new experiences	
11	Accepting the way you are	
12	Taking pain lightly	
13	Listening to others' advice	
14	Keeping your cool in tense situations	
15	Avoiding blaming others for what happened to you	
16	Seeing the glass full rather than half empty	
17	Rebounding from failures	
18	Facing reality and controlling self-deception	
19	Turning resistance into an opportunity	
20	Not being afraid of not knowing what to do	
Total		

Debriefing

If your score is between 150 and 200:

You are managing your emotions effectively and you are able to use them positively. You have emotional empathy.

If your score is between 70 and 149:

You are aware of the power of emotions and have developed a good sense of how useful they can be. However, there is still room for you to improve your ability to manage emotions. Go back and review those items for which you scored 5 or under, and examine your potential for growth.

If your score is between 20 and 69:

Either you do not believe in your ability to manage your emotions or you are missing an opportunity to discover things in yourself that could help you to be a better leader. Reflect on it.

Agenda for Growth
Identify 3 items that you can focus on to enhance your leadership performance with respect to your ability to manage emotions.

Managing other people's emotions

How perceptive are you about other peoples' emotions, both positive and negative?

Keep in mind that emotions are also signals that we are giving to ourselves and others about our value system. You feel happy when your values are impacted positively and unhappy when they are impacted negatively.

As a leader, it is therefore good to listen to people's emotions since they indicate to us what other people care about.

CHAPTER 4

THE SPACE LEADERSHIP COMPETENCY MODEL

9 The Profile of the Space Leader

Drawing on Sergey Ryazanskiy's experiences in his journey towards qualification for space travel to the International Space Station, we established a model of characteristics which constitute a profile of a space leader. This profile combines two categories of characteristics.

Fundamental Characteristics

Form the foundation of the profile.

These characteristics evolved and matured during Sergey's formative years while growing up and were influenced, and encouraged, by his experiences in family, school and university.

Developmental Characteristics

Extend the profile further to incorporate characteristics that were necessary in coping with the challenges of his preparations as an astronaut candidate.

The Critical Skill Set

In assessing Sergey Ryazanskiy's experiences while in space at the International Space Station, we identified the key leadership skills necessary for a successful space mission and built the skills and abilities dimension of the competency model for the space leader.

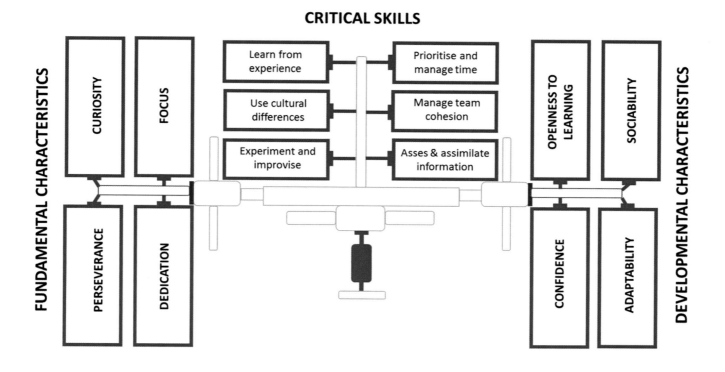

CRITICAL SKILLS

IO Assessment of the Model

This flash survey is a preliminary attempt to assess the quality and relevance of the Space Leadership Model we have presented in this book. With the sponsorship and encouragement of the Moscow School of Management, (SKOLKOVO), we canvassed a group of carefully selected young Russian leaders. We wanted to:

- Know if the key elements of the model make sense to the new generation of aspiring leaders;
- Gather some feedback for future improvements;

- Highlight some of the learnings for our work in developing leaders of the future both in business schools and other educational Institutions.

The results of our assessment exercise reveal a number of interesting issues which would challenge researchers in the international community to explore further and determine the validity of the model in the wider leadership context.

The survey involved 50 future business leaders participating in the MBA program at SKOLKOVO, Moscow School of Management, who were asked to respond to the following questions in an online survey:

Using the following scale as a guide: (1= Not so good; 5=Good; 10=Very good)

According to your own personal leadership experience, do you think that Russian leaders are good at:

1. Being curious about things and people?
2. Focusing on priorities?
3. Being committed and dedicated?
4. Sticking to their goals and strategies?
5. Being open to new ideas and learning?
6. Acting in flexible ways?
7. Leading with confidence?
8. Paying attention to people's needs and expectations?
9. Learning from experience?
10. Managing cultural differences?
11. Trying new things and improvising?
12. Leading teams with harmony and cohesion in mind?
13. Setting up clear priorities?
14. Managing time effectively?
15. Collecting information and using it effectively?

In addition to these 15 questions, MBA participants were asked to provide their opinions on the following 3 additional questions:

- Do you think that there is something missing from questionnaire list above?
- What do you think leaders do well, and which has an impact?
- What do you think that leaders do not do not well and which is important?

Quantitative Survey Results

The following diagram provides an overview of the quantitative analysis of survey results.

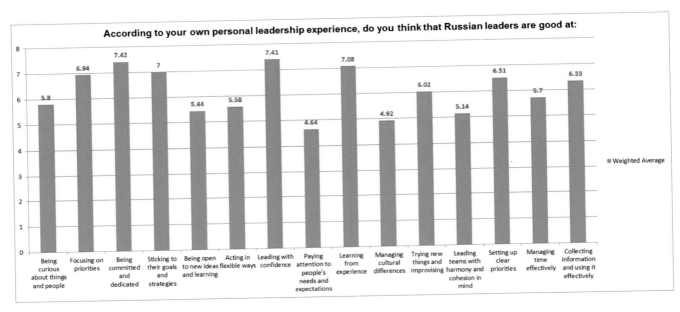

According to your own personal leadership experience, do you think that Russian leaders are good at:

Conclusions

Let's now consider what conclusions can be drawn from our flash survey in respect to the 3 elements of the space leadership competency model.

The Fundamental Characteristics

A noticeable number of respondents consider Russian leaders to be strongest when it comes to being "focused", particularly when it comes to priorities. It's also interesting to note that when it comes to curiosity, respondents consider Russian leaders to be neither in the extreme "good" or "bad", but somewhere in the middle when it came to curiosity.

The Developmental Characteristics

It is very clear that a significant number of respondents consider Russian leaders to be excellent when it comes to leading with confidence, whilst a noticeable number of respondents rate Russian leaders as "less than somewhat good" when it comes to paying attention to people's needs and expectations. This corresponds to sociability in the developmental characteristics of the competency model.

The Critical Skill Set

In their assessment of the performance of Russian leaders with respect to the critical skills of the space leadership competency model, respondents, on average, rank the skills set, as follows, (where 1 reflects where leaders are strongest and 7 where they are weakest):

1. Learning from experience;
2. Setting up clear priorities;
3. Collecting information and using it effectively;
4. Trying new things and improvising;
5. Managing time effectively;
6. Leading teams with cohesion in mind;
7. Managing cultural differences.

Final Comments

"I see earth. It is so beautiful"

Yuri Gagarin

"That's one small step for a man..."

Neil Armstrong

Some scientists claim that if the speed of earth circling around the sun would increase by a factor of 1.4, our planet would escape from the solar system and fly away in space, forever. In other words, our survival as a species is fragile and unpredictable. Who knows what will happen tomorrow? The International Space Station is one attempt to ensure the future survival of our species. We cannot stop from exploring and probing in all fields. We must go on examining our assumptions about space, challenge them and, from a leadership perspective, invent new assumptions which are more in line with the world in the making.

This book is a good example of the ongoing search to have a better idea of where we are and where we are going as living beings. It is about the experience of one man, who has devoted a great part of his life to prepare for and travel to a space station, to conduct research, both alone and in cooperation with others, lead an international team of experts and share some his key leadership learnings. It is a unique leadership experience and part of the ongoing path of exploration that humankind is undertaking, to better understand our existence, and our place in the expansive, known universe. For many people believe that we may, one day, have to move off our planet and colonize far off worlds.

This project can be seen as a modest contribution to the ongoing exploration of space, the final frontier, viewed through the prism of leadership as we understand it. Indeed, it is a first step (following on from the first human being who voyaged into space and man's first step on the moon), that requires crucial leadership qualities, which we have tried to formulate into the two first parts of the book. The leadership model presented in this book, is the result of an attempt to pull some of the learnings together. The main idea is to capitalize on the leadership experience of the astronaut, Sergey Ryazanskiy, and offer his knowledge to all executives and also potential leaders of the world. We are confident that Sergey's story will also appeal to young men and women who will provide leadership in the future.

The International Space Station is a first step for human kind: Testing of the requirements for future of space exploration. Through Sergey's story we get a sense of what team leadership will look like in the new future. This is not science fiction - it is happening up there in space.

This book is a testimony and it is up to the readers to let their imagination run free and extrapolate. It is about one of the doors one must open to move on into the new era of space exploration. The leadership learnings explored in this book are nothing new, but they can help rethink and contribute to the re-invention of the role of the leaders on earth. This book has one basic objective: Encourage as many people as possible to reflect on the core leadership assumptions and values, using the International Space Station adventure as a frame of reference.

Sergey Ryazanskiy and his team on earth

Moscow. October 4, 2019

Suggested readings and tentative pointers for the readers who are just curious

Indicative list of readings and pointers

1. Books

- Hadfield C., An Astronaut's Guide to Life on Earth

- *Davenport C., The space Baron*

- *Launius R.D., The Smithsonian history of space exploration*

- *Asimov I., The last question*

- *Sagan C., Cosmos*

- *Reynolds A., Revelation space*

2- *Movies*

- *Interstellar*

- *2001: A space odyssey and 2010:Odyssey two*

- *Space Rocket „Salyut 7"*

- *Apollo 13*

3. *Innovation (to keep in mind)*

- *Google and the „Quantum Computer"*

- *Russia planning to send a humanoid robot to ISS*

4. *Social and cultural Issues (connected with the space exploration)*

- *Harari Y.N., 21 lessons for the 21st century*

- *Russell S., Human compatible*

- *Sagan C., Billions and Billions*

5. *Web Sites*

- *www.Russianspaceweb.com*

- *www.astronomynow.com*

- *www.space.com*

- *www.solarsystem.nasa.gov*

- *www.chinaspace.info*

Printed in the United States
By Bookmasters